DOWNHILL ALL THE WAY
AN AUTOBIOGRAPHY
OF THE YEARS
1919 to 1939

By the same author

History and Politics
INTERNATIONAL GOVERNMENT
EMPIRE AND COMMERCE IN AFRICA
CO-OPERATION AND THE FUTURE OF INDUSTRY
SOCIALISM AND CO-OPERATION
FEAR AND POLITICS
IMPERIALISM AND CIVILIZATION
AFTER THE DELUGE VOL. I
AFTER THE DELUGE VOL II
QUACK, QUACK!
PRINCIPIA POLITICA
BARBARIANS AT THE GATE
THE WAR FOR PEACE

Criticism
HUNTING THE HIGHBROW
ESSAYS ON LITERATURE, HISTORY AND POLITICS

Fiction
THE VILLAGE IN THE JUNGLE
STORIES OF THE EAST
THE WISE VIRGINS

Drama
THE HOTEL

Autobiography
SOWING: AN AUTOBIOGRAPHY OF THE YEARS 1880 TO 1904
GROWING: AN AUTOBIOGRAPHY OF THE YEARS 1904 TO 1911
BEGINNING AGAIN: AN AUTOBIOGRAPHY OF THE YEARS 1911 TO 1918

The author and Sally

Downhill All the Way

AN AUTOBIOGRAPHY
OF THE YEARS 1919 TO 1939

Leonard Woolf

HBJ

A Harvest/HBJ Book

HARCOURT BRACE JOVANOVICH, PUBLISHERS

San Diego   New York   London

Requests for permission to make copies of any
part of the work should be mailed to:
Copyrights and Permissions Department,
Harcourt Brace Jovanovich, Publishers,
Orlando, Florida 32887.

Library of Congress Cataloging in Publication Data
Woolf, Leonard Sidney, 1880–1969.
    Downhill all the way.
    Continues Beginning again; an autobiography of the years
1911 to 1918. Continued by The journey not the arrival matters;
an autobiography of the years 1939 to 1969.
    Includes index.
    1. Woolf, Leonard Sidney, 1880-1969. I. Title.
JA94.W6A27   1975   320′.092′4   [B]   75-9821
ISBN 0-15-626145-6 (pbk.)

Printed in the United States of America

First Harvest edition 1975

BCDEF

CONTENTS

ILLUSTRATIONS

Frontispiece
The author and Sally

Between pages 128 and 129
Virginia
Monks House, Rodmell
Virginia and Lytton Strachey
(By permission of Mrs. Igor Vinogradoff)
The author and John Lehmann at Monks House
Rodmell Village
The author, Sally, and Virginia in Tavistock Square
T. S. Eliot at Monks House
Virginia and Dadie Rylands at Monks House
Mitz in Rome
Mitz at Monks House
Mitz and Pinka
Maynard Keynes and Kingsley Martin at Rodmell
The author and Nehru
Group at Sissinghurst
The author at Monks House

The herd ran violently down a steep place into the lake, and were choked.

St Luke, Chapter 8

Chapter One

PEACE IN OUR TIME, O LORD

AT THE end of the third volume of my autobiography the
great war of 1914 to 1918 had just ended, having in its
four years killed 10 million men and caused 36 million
casualties. It has been estimated that the direct cost of the
war was about £60,000 million and its indirect cost about
£50,000 million. It destroyed, I think, the bases of European
civilization. We, like everyone who lived through those years,
had been profoundly influenced by them. When the maroons
boomed on November 11, 1918, we were no longer the same
people who, on August 4, 1914, heard with amazed despair
that the guns had begun to boom. In 1914 in the background
of one's life and one's mind there were light and hope; by
1918 one had unconsciously accepted a perpetual public
menace and darkness and had admitted into the privacy
of one's mind or soul an iron fatalistic acquiescence in in-
security and barbarism. There was nothing to be done about
it, and so, as I recorded, Virginia and I celebrated the end
of a civilization and the beginning of peace by sitting in the
lovely, panelled room in Hogarth House, Richmond, which
had been built almost exactly 200 years before as the country
house of Lord Suffield, and eating, almost sacramentally, some
small bars of chocolate cream.

The last sentence seduces me into a digression, though
what I am about to say is not really irrelevant on the first
page of a volume of autobiography, for it is concerned with
the impossibility of telling the truth, the extraordinary difficulty of unearthing facts. The moment one begins to investigate the truth of the simplest facts which one has accepted

9

as true—about one's own life, for instance—it is as though one had stepped off a firm narrow path into a bog or quicksand—every step one takes one sinks deeper into the bog of uncertainty.

For instance, the above statement about Hogarth House is not true, though for years I believed it to be true. When in 1915 we took a lease of Hogarth House, it was part of a large eighteenth-century mansion which belonged to a lady living in Bushey. The mansion had been very ingeniously divided into two houses, one called Suffield House and the other Hogarth House. We were told at the time that the whole house had been built originally in 1720 as a country house for Lord Suffield and had been converted into two houses with two front doors in the nineteenth century. That it was built in 1720 was, I think, almost certainly true. That was, it seems to me, about the best moment for English architecture, at any rate for the medium-sized, aristocratic country house like the one which Lord Suffield did not build in Richmond in 1720. The interior of the original undivided house must have been perfect. All the rooms were panelled, the ones on the ground floor with a certain amount of chaste ornamentation; in the others the panels became progressively plainer as one went up from floor to floor. Every room was beautifully proportioned. Most houses and gardens are, like most of the people who make them or live in them, featureless, amorphous; the houses are closed boxes in which people live, the gardens open boxes in which people grow flowers or vegetables. Occasionally one comes across a house upon which those who built it or lived in it have imposed a character and form markedly and specifically its own, as though it were a person or a work of art. Hogarth House was one of these. All the rooms, even when we first saw them in the dirty, dusty desolation of an empty house, had beauty, repose, peace, and yet life. One felt at once that each of them only needed a table and chair, a bed or a bookcase to become the perfect cell in which a human being might eat and sleep, talk, read, or work. Perhaps the people who for 200 years had been doing just that in these rooms had left the aura of their lives in them, but more prosaically it was matter—bricks and mortar and wood—and the way in which they had been used 200 years before which gave to Hogarth House its extraordinary character of being the perfect envelope for everyday life. It was partly its combination of immense solidity with grace, lightness, and beauty. The electrician who had to take a wire through the inside wall of the drawing-room, told us that in all his experience he had never seen as thick an inside wall in a house. In the room itself one felt the security from anything like a hostile world, the peace and quiet, in this tremendous solidity of walls, doors, and windows, and yet nothing could have been more light and graceful, more delicately and beautifully proportioned than the room itself, its fireplace and great windows, its panelling and carved woodwork.

In the house before it was divided there was a great hall and a very beautiful staircase, and on the first floor a tremendously broad corridor. We lived in this house from 1915 to 1924. After the war, in 1920, the owner refused to renew our lease, but offered to sell us the whole property, i.e. both Suffield and Hogarth House. We bought it for £2,000, and at one moment we thought we might restore the two houses to their original condition as one great country house and live for the remainder of our lives in this magnificent Suffield House. But it was a project half serious and half a day dream. By 1924 we had abandoned it, for the house would have been much too large for us and we had decided that it was time to move back into London. So we sold what had once been Suffield House and moved to Tavistock Square.

But had it ever been Suffield House? Had it ever belonged to Lord Suffield? It is impossible to know, but what is quite certain is that no Lord Suffield built it as his country house in 1720, because the barony of Suffield was created in 1786. So much for the truth about the genealogy of a house. I recently discovered that when I bought Monks House, Rodmell, in 1919, what I was then told about its name and its genealogy was also quite untrue. As I recorded in a previous volume of my autobiography,[1] it was said to have been called Monks House, because in the fifteenth century it belonged to Lewes Priory and the monks used it for their 'retreats'. I said that I hoped the story was true, but I rather doubted such legends about houses when there is no documentary evidence for them. Since writing that, I have examined the deeds for Monks House and find that the story is entirely untrue. The deeds go back to 1707 and record the names of everyone who lived in it (together usually with the names of all their sons and daughters) from 1707 to 1919, when I bought it from the heirs of Jacob Verrall. From 1707 to 1919 only three families owned and lived in it. In 1707 John Cleere or Clear, carpenter of Rottingdean, acquired it from James de la Chambre. From 1707 to 1796 it remained in the Clear family, passing from father to son. In 1779 John Clear, carpenter, great-grandson of the original John Clear, carpenter, inherited it, and left it, when he died in 1782, to his son James. (He left five guineas to each of his two sons, Edward and Thomas, and a guinea to his granddaughter Charity.) In 1796 James Clear sold it to John Glazebrook. All this time, from 1707 to 1796, the house was called Clear's.

From 1796 to 1877 the house was in the Glazebrook family, 33 years in the hands of John Glazebrook and 48 years in those of his widow, Mercy, and his son William,

1 *Beginning Again*, p. 61.

except for a short interval in 1829 when it was sold for £300 to Matthew Lower, publican, of Rodmell, who resold it for £300 back to William Glazebrook. The Glazebrooks must have been connected with the Clears and with the house already 30 years before they actually acquired it, for in a mortgage deed of 1765 the property is described as a messuage in the tenure of John Clear and John Glazebrook. The Glazebrooks were millers; John Glazebrook in 1749 bought the mill which was up on the down above Rodmell, and it remained in the family for 128 years, when the executors of William Glazebrook sold it to Jacob Verrall in 1877.

At the same time, in the same year, the executors also sold what is now called Monks House to Jacob Verrall. From 1796 to 1877 the house was called Glazebrook's except for a short interval when it was called Lower's. The first time that it was ever called Monks House in any document was when it was advertised for sale in 1919 on Jacob Verrall's death. The story that Monks House belonged to the Lewes monks in the fifteenth century is therefore just as false as the story that Suffield House was built by Lord Suffield in 1720.

It is rather depressing for an autobiographer starting on a fourth volume to find in this way that most of his facts are will-o'-the-wisps and that it is almost impossible to tell the truth. Facts about the houses in which one lives during the whole journey from the womb to the grave are not unimportant. The house—in which I include its material and spiritual environment—has an immense influence upon its inhabitants. Looking back over one's life, one sees it divided by events into compartments or chronological sections, e.g. I was at St Paul's School 1894 to 1899, I was in the Ceylon Civil Service 1904 to 1911, I lived through the war of 1914 to 1918 and the war of 1939 to 1945, I was married in 1912. All such momentous or catastrophic events moulded the form of one's life, disrupted or distorted its movement. But what has the deepest and most permanent effect upon oneself and one's way of living is the house in which one lives. The house determines the day-to-day, hour-to-hour, minute-to-minute quality, colour, atmosphere, pace of one's life; it is the framework of what one does, of what one can do, and of one's relations with people. The Leonard and Virginia who lived in Hogarth House, Richmond, from 1915 to 1924 were not the same people who lived in 52 Tavistock Square from 1924 to 1939; the Leonard and Virginia who lived in Asham House from 1912 to 1919 were not the same people who lived in Monks House from 1919 to 1941. In each case the most powerful moulder of them and of their lives was the house in which they lived. That is why looking back over my life I tend to see it divided into sections which are determined by the houses in which I lived, not by school, university, work, marriage, death, division, or war.

When I bought Monks House in 1919 there was an auction of all its contents. It took place in the garden on a marvellous sunny summer day. All the village attended, including descendants of the long line of Glazebrooks who had already been moving about in the house and garden 155 years before. 1919 was a great fruit year in Sussex; the trees were laden with plums and pears and apples. The branches of an enormous apple-tree heavy with great red apples hung over the yew hedge along which we stood bidding or just watching the auction; and every now and again someone would pull a great red apple off the tree and eat it. There is nearly always something sad and sinister in the auction of the contents of a house, a kind of indecent exposure of the lives of dead men, women, and children. This is particularly the case when the auctioneer reaches those cold and comfortless attics in which in distant days servants slept on iron bedsteads. As the auctioneer's men carried the furniture, glass, and china, ornaments, pictures, and the accumulated odds and ends of a family's possessions on to the Monks House lawn, it seemed at moments as though one were watching the disembowelling, not merely of a house, but of time. Old Jacob Verrall's[1] wife Lydia was, I think, a connection of the Glazebrooks and much of the furniture etc. must have belonged to them and to have been in the house for a century and more. Some of the old furniture and china was beautiful and was bought up at quite high prices by dealers. I bought three pictures painted on wood by a Glazebrook in the middle of the nineteenth century or perhaps a little earlier. They were painted in that curious stiff uncompromising style of the inn signboard of a hundred years ago. One was of a middle-aged man, very dark and bewhiskered, and another of a man holding a horse. The third was of four children heavily swaddled in hats and coats standing stiffly in a line in front of the house. They were, I am sure, the Glazebrook children of a hundred years ago. Their spirits, I almost felt and feel, walk in the house, clattering up and down the narrow stairs, now deeply worn by the countless comings and goings of Clears, Glazebrooks, and Verralls. At the top of the stairs you can see the place where they had once put a small gate to prevent the children plunging downstairs. And once when a floorboard was taken up by a workman we found a tiny little wooden eighteenth-century shoe; another time I found in the cellar a George III fourpenny piece which appeared to have been charred in a fire.

These little facts are not, I think, either unimportant or irrelevant. In the atmosphere of both houses, Monks House in Rodmell and Hogarth House in Richmond, there was something similar. In both one felt a quiet continuity of

[1] I gave some facts about old Verrall and his wife in *Beginning Again*, pp. 63 and 64.

people living. Unconsciously one was absorbed into this procession of men, women, and children who since 1600 or 1700 sat in the panelled rooms, clattered up and down stairs, and had planted the great Blenheim apple-tree or the ancient fig-tree. One became a part of history and of a civilization by continuing in the line of all their lives. And there was something curiously stable and peaceful in the civilization of these two houses. In 1919 when we bought Monks House, Virginia was only just recovered or recovering from the mental breakdown which I have described in *Beginning Again*; in 1919 we still had six years of life in Hogarth House before we moved into London. Those six years were, I am sure, crucial for the stabilizing of her mind and health and for her work, and I am quite sure that the tranquil atmosphere of these two houses, which was in their walls and windows and gardens and orchard, but also in the soothing, chastening feeling of that long line of quiet people who century after century had lived and died in them—I am sure that this tranquil atmosphere helped to tranquillize her mind.

At the end of 1919, then, we were the owners of two houses. Our expenses during the previous twelve months had been £845. We had been printing and publishing books in the Hogarth Press for two years, but the Press was still a hobby which we practised in our spare time. The three books which we published in 1919 were Virginia's *Kew Gardens*, T. S. Eliot's *Poems*, and J. Middleton Murry's *Critic in Judgment* —on which we made a net profit of £26, 3s. 10d. Virginia had just begun her 'career' of a novelist. Her first novel, *The Voyage Out*, had been published by Duckworth four years ago in 1915. Her second novel, *Night and Day*, had just been published, also by Duckworth, in October 1919. *The Voyage Out* had received high praise, and so did *Night and Day*, but to a less degree. Neither book was a success financially either for the author or publishers, for, as I have recorded elsewhere, nine years after it was published Duckworth had sold only 2,238 copies of *Night and Day*. Virginia's only other publications by the end of 1919 were *The Mark on the Wall*, which the Hogarth Press published in 1917, and *Kew Gardens*, which we published in May 1919. She did not begin to write her third novel, *Jacob's Room*, until April 1920, but she wrote some short pieces like *An Unwritten Novel* and did a certain amount of reviewing in the *Times Literary Supplement* and *Athenaeum*. Her earnings in 1919 from her writing, at the age of 37, were £153, 17s. 0d. Virginia was 40 years old before she earned a living wage by writing; if she had had to earn her living during those years, it is highly improbable that she would ever have written a novel.

By 1920 I had accumulated a considerable number of paid and unpaid occupations. I was editor of the *International Review* on a salary of £250. I did a good deal of freelance journalism, mostly for the *New Statesman*, but a certain amount for the *Nation* and the *Athenaeum*. I earned in 1919 £578, £262 by freelance journalism, £250 from my editorship, and £66 from my books. The *International Review* was a monthly financed by the Rowntrees; I had an office in Red Lion Court, Fleet Street, in which sat Miss Matthaei, Assistant Editor, and Miss Green, Secretary. I went to the office three or four days a week. We did a great deal of work. My idea was that the *Review* should cover the whole field of foreign affairs, international relations, and the problem of preventing war which centred in the inchoate League of Nations. My main object was to try to put before readers the facts without a knowledge of which it was impossible even to begin to understand the intricate problems of the international chaos created by the war. I therefore had two features in the paper which I thought of great importance. The first, which I wrote myself, was an 'international diary'; in it I dealt with the chief international events of the previous month. In order to produce this diary Miss Matthaei and I read French, German, Austrian, Italian, and Spanish daily papers. The second feature was, I think, something quite new in this kind of journalism. I had a section called 'The World of Nations: Facts and Documents'. It ran to 30 or 40 pages and it contained all kinds of documents, most of which were unobtainable elsewhere. We took any amount of trouble to obtain documents, and the kind of thing we published is shown by the contents of a single number, November 1919: (1) Message of Admiral Kolchak to the peoples of Siberia and instructions to military officers; (2) Declaration of the Ukrainian Government regarding Denikin; (3) The text of an Anglo-Persian Treaty; (4) The text of an alleged treaty between Germany and Japan which had been published in America but not in Britain; (5) A translation of the full text of the new German Constitution.

In my search for documents I had some curious experiences; the following was one of the most interesting. I do not remember how or when I first got to know Theodore Rothstein, a Russian Jew living in London. He is frequently mentioned in the diaries of Wilfrid Scawen Blunt, for in 1907 he was London Correspondent of the *Egyptian Standard* and worked closely with Blunt and Brailsford for Egyptian independence. When I knew him in 1919 he was unofficial ambassador of the unrecognized Bolshevik Government. He told me that, when the Bolsheviks first seized power, the London police arrested him and put him on a ship lying in the Pool just below London Bridge, meaning to deport him in it to Russia. Rothstein knew Lloyd George and had had 'off the record' communication with him on behalf of Lenin. He succeeded in getting a letter to the Prime Minister smuggled out of the ship, and orders were immediately given to the police to release the Russian 'ambassador'.

Rothstein was a short, stumpy, bearded, bespectacled

revolutionary who looked like Karl Marx. He was the first of the many hundred per cent. dyed in the wool, dedicated communists that I have had the misfortune to come across in the last 45 years of my life. Communists, Roman Catholics, Rosicrucians, Adventists, and all those sects which ferociously maintain as divine or absolute truth, monopolistically revealed to them, an elaborate abracadabra of dogmas and fantasies, fill me with melancholic misery. The ruthlessness and the absurdity of the believers' beliefs reduce me to despair. What is the point, one feels, of any political, social, scientific, or intellectual activity if civilized people in the twentieth century not only accept as divine truth the myths dreamed by Palestinian Jews two or three thousand years ago or by German Jews a hundred years ago, but also condemn to Hell, death, or Siberia those who disagree with them?

Rothstein, as I said, was the first of these modern civilized savages, these communist fanatics, that I came across. Outside the circle of his Marxist religion he seemed to me a nice man and highly intelligent; inside the magic circle he was a cross between a schoolman and a dancing dervish. He would expound the gospel of Marxism-Leninism to me at great length in that dreadful jargon of meaningless abstractions which has become the language of communism and the excuse for the torture or killing of hundreds of thousands of human beings. Some time in 1919 he came to me and said that he had the full text of a number of speeches made by Lenin since his return to Russia. None of these very important speeches and statements of policy had been reported in the British or American press, and he was willing to give me translations of them if I would publish them in the *International Review*. I said that I would, and then I had my first experience of the behaviour of the real underground revolutionary.

The question was how the typescript of the translation of

Lenin's speeches should be physically handed over by Roth-stein, his agent, to me, the editor. Having had no experience of revolutionaries, secret agents, or spies, I naturally thought that it would be sent to me in the ordinary way through the post. Rothstein was horrified at such a crude and naïve idea. There was, he said, and in this he was correct, still operating a censorship, which was a legacy from the war, and if the authorities knew of the existence of verbatim translations of Lenin's speeches, they would refuse to allow publication; we must on no account allow the police to know that he was going to give them to me. The only way to defeat the police was for me to follow his instructions meticulously. On Wednesday afternoon I was to walk down the Strand to-wards Fleet Street, timing it so that I should pass under the clock at the Law Courts precisely at 2.30. I must walk on the inside of the pavement and precisely at 2.30 I would meet Rothstein under the clock walking from Fleet Street to Trafalgar Square on the outside of the pavement. He would be carrying in his right hand an envelope containing Lenin's speeches, and, as we passed, without speaking or looking at each other, he would transfer the envelope from his right hand to mine.

This elaborate procedure was carried out and I sent the speeches to the printer to be printed in the next issue of the *International Review*. I do not know how the police dis-covered that we were going to publish these documents or why the authorities thought that it would be dangerous for the British people to know what Lenin was saying—it seems rather fantastic to believe that a Secret Service man was always trailing Rothstein and saw him hand over the enve-lope to me outside the Law Courts. At any rate a few days later the police went to the printers, seized the documents and, I think, some type which had already been set, and forbade publication.

The British public were thus prevented from knowing what Lenin was saying in Russia at the historical moment when that knowledge was most interesting and politically important. This is one of the many instances of congenital stupidity in secret services and censorship which I have come across in my life and which, however often I come across them, fill me with innocent surprise. No intelligent person who went about his business in London in 1918 and 1919, who talked to the common man and knew what is called the climate of opinion, could possibly have thought that the number of people who would have been politically influenced by reading Lenin's speeches in the *International Review* would have exceeded the number of righteous men whom Abraham and the Lord found in Sodom. But once one begins to try to suppress some knowledge or some opinions, one loses all sense of proportion and relevance in one's obsession with the danger of ideas. In the end the only safe course for the worried, nervous policeman and the cloistered censor, sitting aloof in his office with the blue pencil in his hand, is to try to suppress all knowledge and all thought.

I do not think that I saw Theodore Rothstein many times after the fiasco with Lenin's speeches. It was not very long before he went back to Russia. I was told—I do not remember by whom or know with what truth—that he became a Commissar in Samarkand or some other remote province of the Soviet Empire. He was such a ruthless dogmatist and such a dedicated Leninist-Marxist that he must, I think, sooner or later have been liquidated in one of the great purges by some equally dedicated and ruthless comrade.

To someone like myself born in the comparative civilization of the nineteenth century one of the horrors of life since 1920 is its senseless savagery. If one shuts one's eyes or one's mind, it is just possible to ignore the millions of Jews slaughtered in Hitler's gas chambers and the millions of unstigmatized persons killed in concentration camps and 'on the field of battle' during the 1939 war. But somehow or other the crowning point of barbarism seems to have been reached in the kind of doctrinal or racial cannibalism that has swept over the earth. The merciless savagery with which Spaniard treated Spaniard in the civil war, or Italian treated Italian under Mussolini, or German treated German under Hitler, or African is now treating African in the Congo, makes such outbreaks as the Armenian atrocities, which horrified Gladstonian liberals towards the end of the nineteenth century, appear insignificant. 'Dog does not eat dog' and 'a wolf does not make war on a wolf' are such ancient truths that they are proverbial, but in the twentieth century large-scale fratricide has been common among patriots, monarchists, republicans, fascists, nazis, socialists, anarchists in Germany, Italy, Spain, and Africa. But the doctrinal cannibalism of communists since 1917, particularly in Russia, has been even more repulsive if only because of its scale. The liquidation in 1930 of the Russian kulaks— peasants numbering with their families five million persons —is one of the most dreadful stories in the whole of history.[1] And no one will ever know how many hundreds of thousands of Russians have been liquidated by Russians in the last 40 years—in forced-labour camps, prisons, judicial murders, purges. When one reads that a million kulaks have been ruined or done to death because they were rather prosperous peasants, or 500,000 Russian communists have been killed by Russian communists because they were either right deviationists or left deviationists, or six million German Jews

[1] Sir John Maynard in *The Russian Peasant* says that 'it can only be compared for ruthlessness with the wholesale removals of population by the ancient monarchies, or the expulsion of the Moors from Spain or the Jews from Germany'.

have been killed by German Christians because they were Jews, one cannot feel that each one of these persons was an individual like oneself, that every one of a million Russian peasants when he was suddenly driven out of his house and off his land to starve and die with his family in the snow, and each one of those hundreds of thousands of Russian communists when he felt himself rotting to death in the Siberian labour camp, and each of those six million Jews when he found himself being driven naked by the nazi guards into the gas chamber, suffered, before the final annihilation of death, the same agony which you or I would suffer if it happened to us.

I do not think that to say this is sentimental or here irrelevant. At any rate it has, I know, to me personally a peculiar and profound relevance. I have known as individuals and friends in London two Russians who went back to Russia and put their heads into the noose of Stalinist communism. I feel pretty certain that in each case the noose was pulled sooner or later and my friends were liquidated. If you had searched the world, you could not anywhere have found two men more unlike each other than Theodore Rothstein and Prince Mirsky, the Russian Jew and the Russian aristocrat. Rothstein was, as I have said, a mediaeval schoolman born into the twentieth century, a pedant and fanatic where the gospel of Karl Marx was concerned. He was, I think, by nature a gentle and civilized man, who loved talk and the intellectual pleasure to be derived from the intricate working of good brains. But he had been caught in the cruel inhuman machinery of communism. If he were not ruthlessly liquidated, he would himself have been a ruthless liquidator.

Let me leave Rothstein for the moment being shot by a comrade or shooting a comrade in one of those purges by which behind the iron curtain men build the perfect society. Let me turn to Prince Mirsky. Mirsky was a stranger man than Rothstein. I always felt that he was fundamentally one

of those unpredictable nineteenth-century Russian aristocrats whom one meets in Aksakov, Tolstoy, and Turgenev.
Sometimes when one caught in a certain light the vision of
his mouth and jaw, it gave one that tiny little clutch of fear
in the heart. It made one think of Turgenev's mother flogg-
ing the servant to death. I have known only a very few
people with this kind of mouth; its sinister shape comes, I
think, from the form of the jaw and arrangement of the
teeth. There is always the shadow of a smile in it, but it is
the baleful smile of the shark or crocodile.[1] Mirsky had this
kind of smile. It may have had no psychological significance
and he may well have had nothing cruel or sharklike in his
character. In all our relations with him he seemed an un-
usually courteous and even gentle man, highly intelligent,
cultivated, devoted to the arts, and a good literary critic.
He had, at the same time, that air of profound pessimism
which seemed to be characteristic of intellectual Russians,
both within and without the pages of Dostoevsky. Certainly
Prince Mirsky would have found himself spiritually at home
in *The Possessed* or *The Idiot*.

One day Mirsky came to us in Tavistock Square and told
us that he was going back to Russia. This must have been
in 1931. By that time one knew something of the kind of
life (or death) that an intellectual might expect in the Russia
of Stalin. It seemed madness, if not suicide, for a man like
Mirsky voluntarily to return to Russia and put himself in the

[1] One day when I was travelling by train along the south coast of
Ceylon from Matara to Galle, on the platform of one of the stations
through which we passed there were dozens of dead sharks. I had
never seen anything like it in Ceylon and I do not know why they
were there. On each dead face there was this sinister grin. Talking to
Mirsky in a London sitting-room, as he suddenly turned his head to say
something and there was a glint of teeth and smile, I was back in Ceylon
twelve years ago in the railway carriage looking at the rows of dead,
smiling sharks.

power of the ferocious fanatics who could not possibly have the slightest sympathy with or for him. We knew Mirsky well enough to say so. He was extremely reticent, shrugging it all off with some platitude, but he left us with the impression of an unhappy man who, with his eyes open, was going not half, but the whole, way to meet a nasty fate. We never saw him again.

The fate of Mirsky and of Rothstein seems to me terribly typical of our time. Both of them, as I said, were almost certainly liquidated, which means that they were in some horrible way put to death, murdered. Even if they were not, they must have escaped by some accident, for thousands of men like them have been liquidated in Russia. Contemplating this and them, I feel the horror of the savagery of contemporary man in a way in which I do not feel it when I hear of the more horrible stories of the massacre of millions. I knew them as individuals, and it is as an individual that I feel their fate, this liquidation, this senseless torture and killing of two harmless individual human beings. For what after all could be more harmless than the slightly ridiculous bespectacled Rothstein spinning the endless web of the Marxian abracadabra or Mirsky endlessly discussing the magnificent absurdity of Tolstoy or the niceties in the torrential style of Dostoevsky? That Theodore Rothstein may have been himself potentially as cold-blooded a murderer as his cold-blooded murderers, or that Prince Mirsky may have been potentially as inhumanly cruel as so many other Russian aristocrats, does not contradict or make nonsense of what I have just written; it only underlines the senseless political and social stupidity of contemporary Europe. I have sat talking in Richmond with Rothstein and in Paris and Tavistock Square with Mirsky, and I know that what interested them and what gave them pleasure were things of the intellect and the arts, painting, music, and literature.

In a world which had the slightest claim to civilization, they would have lived and died civilized men, doing or suffering no public evil. As it was, their lives became hopelessly entangled in the wheels of an idiotic, barbarous social and political system, and the misery and death which they suffered (or which perhaps they caused) were inflicted on pretexts or for reasons which have no sense, no reality, no importance for the vast majority of the human race. Power and the struggle for power are of course realities involved in the machinery of communism and Soviet Russia in which Rothstein and Mirsky became fatally involved; but power is always the concern of a tiny minority. The Rothsteins and Mirskys and the thousands of anonymous victims of communism are sacrificed for words and phrases, tales 'told by an idiot, full of sound and fury, signifying nothing'.

I know that I am prejudiced against communism, which seems to me in some ways worse than nazism and fascism. *Corruptio optimi pessima*—the greatest evil is the good corrupted. The Hitlers and Mussolinis are just thugs or psychopaths, savages who in all ages have formed the scum of society; their imitators like Oswald Mosley rouse in me no emotion more serious than contempt. But communism has its roots in some of the finest of human political motives and social aspirations and its corruption is repulsive. The first time I met Mirsky was in Paris, in Jane Harrison's flat. Jane Harrison, the brilliant Newnham classical scholar, was one of the most civilized persons I have ever known. She was also the most charming, humorous, witty, individual human being. When I knew her she was old and frail physically, but she had a mind which remained eternally young. She liked Mirsky and enjoyed talking to him, and he, I felt, sat at her feet. That from that environment he should have been drawn into the spider web of Soviet Russia to be destroyed there fills one with despair, despair that communism, by *corruptio optimi*, again and again and again has 'lighted fools the way to dusty death'.[1]

I have reached the period in my autobiography in which our lives and the lives of everyone have become penetrated, dominated by politics. Happy the country and era—if there can ever have been one—which has no politics. Ever since 1914 in the background of our lives and thoughts has loomed the menace of politics, the canker of public events. (One has ceased to believe that a public event can be anything other than a horror or disaster.) Virginia was the least political animal that has lived since Aristotle invented the definition, though she was not a bit like the Virginia Woolf who appears in many books written by literary critics or autobiographers who did not know her, a frail invalidish lady living in an ivory tower in Bloomsbury and worshipped by a little clique of aesthetes. She was intensely interested in things, people, and events, and, as her new books show, highly sensitive to the atmosphere which surrounded her, whether it was personal, social, or historical. She was therefore the last person who could ignore the political menaces under which we all lived. *A Room of One's Own* and *Three Guineas* are political pamphlets belonging to a long line stretching back to *Vindication of the Rights of Women* by Mary Wollstonecraft, and she took part in the pedestrian operations of the Labour Party and Co-operative Movement. And by 'pedestrian' I mean the grass roots of Labour politics, for she had a branch of the Women's Co-operative Guild meeting regularly in our house in Richmond and we had the Rodmell Labour Party meeting regularly in Monks House, Rodmell.

[1] Since writing the above, I have been told by Malcolm Muggeridge, who saw a good deal of Mirsky in Moscow in 1932-33, that Mirsky just before the war was sent to a camp for ten years and either died or was shot there.

The theme of politics and public events must therefore become more important and more persistent in this autobiography from 1919 onwards. It was not merely that I became more and more actively immersed in them. We lived our daily life and ate our daily bread in the shadow of recurring crises and catastrophes. When peace at last came in 1918, it was, of course, like the break in the appalling sky which must have covered poor Noah's ark, a gleam of sun 'after the end of the hundred and fifty days' when 'the fountains of the deep and the windows of heaven were stopped and the rain from heaven was restrained'. Of course we welcomed the dove with the olive leaf in her beak. We put out a few flags and a few hopes hesitantly, apprehensively. Almost immediately the flags drooped, the olive leaf withered, the hopes faded. In the years 1918 to 1939 one impotently watched a series of events leading step by step to barbarism and war: the Versailles Treaty and the canker of reparations; the creation of Stalin's Russia, the iron curtain, and the cold war; the rise of fascism and nazism; the failure of the League of Nations; the menace of nuclear war; the Hitlerian Götterdämmerung.

In all this gloom the darkest spot seemed to me and to many other people, at any rate until Hitler came to power in 1933, Stalin's Russia. I have described in *Beginning Again* (pp. 207-215) how, when the Tsarist regime fell, we welcomed the 1917 revolution with the same kind of relief and elation which Wordsworth felt in 1789 in the first days of the French revolution. The disillusionment was all the greater. At first one was puzzled by the senselessness of the iron curtain—the shutting off of millions of civilized persons in the twentieth century from the rest of the world and from truth. Then gradually it became clear that the communist rulers of Russia were determined not only to keep their subjects in darkness and ignorance, but also, if possible, to keep the rest of the world in a state of fluid chaos. The foreign policy of the Soviet Government was always simple and consistent: they fished in troubled waters, but they were also continually trying to make the waters troubled so that they could fish. Hence the cold war.

Then still more gradually one became aware of the senseless barbarism of communist society behind the iron curtain. Here again, so far as I was concerned, a realization of the truth only came gradually by personal experience, by knowing some insignificant individual caught and crushed in the inhuman machinery of the Soviet state—which according to Theodore Rothstein and Karl Marx ought to have withered away. I remember the shock of the first time when I caught a glimpse of this monstrous juggernaut crushing a little individual (and innocent) fly. I knew a young woman, whom I will call Jane, who married, in the early years of the Soviet regime, a Russian scientist employed in Russia. They lived in Leningrad and she was allowed to come for a few weeks every year and visit her mother in England. She always came to see me. I knew her well and she used to tell me about her life and her views with the greatest frankness. Jane was intelligent and had that spontaneous, generous political enthusiasm often characteristic of the young, and particularly the female young. She was a communist before she married her Russian, for she was one of the many intelligent young people on the political Left who in the early 1920s were depressed by the dreary record of German social democracy and were carried away by the promises of communists and communism. Every year for a year or two after her marriage the day used to come when Jane would burst into my room in the highest spirits and tell me of all that the communist regime was doing and was going to do for the 'toiling masses'. Eventually when a year had gone round and the day came for her visit, things had changed. She was depressed and worried, and admitted that she was anxious about the way things were going in Russia. According to her account, the idealistic asceticism which Lenin had imposed upon the party was breaking up. What had attracted her in communism and what she had found in Lenin's Russia was the selfless dedication of the leaders to the task of transforming Russia into a society 'in which the free development of each is the condition of the free development of all'. Lenin was a ruthless man, and he created a ruthless party; but he was ruthless with himself and he insisted upon communists being ruthless with themselves. Their aim was socialism pure and undefiled, both in theory and in practice. Nearly all communists whom I have known have been very callow or very cunning. (Some of the most hoary old Marxists, like Rothstein, were really both at the same time.) Jane was as politically callow, when a young woman, as an unfledged sparrow. She married and went to Russia believing that the communists and she with them were out to build Utopia—the New Jerusalem and Cloud-Cuckoo-Land. After Lenin's death and the struggle for power which followed it, even Jane could see that idealism and a good deal of freedom were dying out of communism and the Soviet Republic. Something new had come in with Stalin and Stalin's men. Any clouds or cuckoos faded away, for the new rulers were tough and realists. They drove about in big cars and you had to be careful of what you said about them. Jane went back to Russia depressed and uneasy.

It must have been in 1936 that Jane returned to England from Russia for good and came to see me. She was in tears when she told me her wretched story. Her husband, she said, was a scientist, and a devoted scientist who took no part in politics. He was an extremely cautious man and, whatever he may have thought about the regime, never criticized it. One day he did not return home from his laboratory; he just disappeared. Some time later she received an official notification directing her to take some of his clothes and personal possessions to a certain government building. When she got there, she found a series of what looked like ticket offices each labelled with letters of the alphabet. She was told to hand in her husband's possessions at the ticket office labelled with the initial letter of his name. There were long queues of people, like herself, waiting to hand in bundles and suitcases at the various ticket offices. She was given a receipt for her husband's possessions. She never saw him again; eventually she received a letter from him from a labour camp in the Far East. When the time came for her annual visit to England, she was given her permit. In London she went to see Mirsky and consulted him as to what she should do; he strongly advised her not to return to Russia, and she took his advice.

Thus the enormous machinery of the Soviet state was used to disrupt the lives of these two little innocuous insects, Jane and her husband. And senselessly this two-handed engine at the door smote, and smote no more, so far as these two insects were concerned. It is this streak of senselessness in the savagery of communist, and indeed all authoritarian, states which repels and puzzles one. I am quite sure that Jane was speaking the truth when she said that her husband was entirely non-political and that the only possible reason for his liquidation was that one of his fellow-scientists, working in the same laboratory, who was arrested at the same time, was notoriously indiscreet in his criticism of the regime.

I got from Jane another glimpse of the grotesque nightmare of mutual fear in which, under the shadow of the secret police, both rulers and ruled lived in Russia. 1937 was the centenary of the great Russian poet Pushkin's death. Jane gave me a manuscript translation of a short book by

Pushkin which had not been translated into English before, I think, or at any rate was not in print; as far as I can remember, it was autobiographical and extremely interesting. The translation, which was excellent, was by a Russian woman, a friend of Jane's. The suggestion was that the Hogarth Press should publish it in the centenary year. I was eager to do so, but instantly the menacing spectre of the Soviet Government and the secret police and Siberia rose up out of the manuscript to make us pause in Tavistock Square, London, W.C. 1, in the year 1936. Jane explained the difficulty to me. Whether in fact to publish was a nice question, the nicety being for her friend the thinnest partition between life and death. For it to be known that someone in Russia was the translator of a book written by Pushkin over 100 years ago and now published in London might or might not be extremely dangerous for the translator. Whether it would lead to the liquidation of the translator or not would depend upon the amount of terror and fear obtaining at any particular moment among the rulers, secret police, and the ruled in Leningrad and Moscow. Jane had therefore arranged with her friend that, if I decided that the Hogarth Press would like to publish, I should send her a telegram saying simply: 'Many happy returns'. If it was safe to publish, she would reply: 'Many thanks for good wishes'; if it was not safe, she would not reply. I sent off my wire and got no reply, and so the book was never published. It is interesting to compare this incident with the action of the London police with regard to Lenin's speeches which I have related above. The idea that the publication of a translation of a book by Pushkin in London could have harmed in any way the Russian state or people was as fantastic as the idea that the publication of Lenin's speeches in the *International Review* could have done the slightest damage to the British state or people. But, as I said before, censorship of thought and opinion in the hands of a government and its police is a malignant canker which grows and grows, gradually destroying its environment, the mind of society. If there are dangerous thoughts, all thought may be dangerous; it is safer therefore to suppress as much as you can whether it be Lenin or Pushkin.[1]

The insistent pressure of politics, increasing rapidly as soon as war ended, caused me to stand rather halfheartedly for Parliament. It began in 1920. In those days there was a Combined English University Constituency which included all the English universities other than Cambridge and Oxford; they elected two M.P.s. After the khaki election of December 1918, the Seven Universities' Democratic Association asked me whether I would consider becoming a candidate at the next election. It was not a prospect which filled me with any enthusiasm. As a secretary of the Labour Party Advisory Committees on International and Imperial Affairs, which after a time took to meeting in a Committee Room in the House of Commons, I got to know a good many Labour M.P.s who were members of my committees, and I saw from the inside the kind of life they had to lead. The hour-to-hour and day-to-day life of professional and business men in their offices and at their 'work' consists largely of time wasted in a vicious circle of unnecessary inaction or futile conversations. Nearly all important business is done effectively and expeditiously outside the office, which can be reserved mainly as a place in which one dictates and signs letters. (That is why during the last 50 years, whether as editor or publisher, I have always stipulated that I would spend the minimum amount of time 'in the office'.) The business life of a backbench M.P. in the 1920s seemed to me the acme of futility and boredom. He was, no doubt, a member of what was said to be the best club in London,

[1] I suppose that the secret police were not merely afraid of Pushkin, but afraid that a Soviet citizen should know a British citizen.

but he had to be perpetually in it, endlessly doing nothing as he waited to record his vote at the next division.

This prospect of joining the melancholy procession of backbenchers through the lobbies of the House of Commons, as I said, did not appeal to me, and for some time I hesitated to become a candidate. However, eventually I agreed, and in May 1920 was adopted as a candidate. I agreed partly, and rather pusillanimously, because there was really no chance of my being elected. The sitting members were the Conservative Sir Martin Conway and the Liberal Herbert Fisher. I must admit that a second reason which induced me to stand was the prospect of standing against Herbert Fisher. Herbert was a first cousin of Virginia's, a man of great charm, both physical and mental, but also the kind of man whom in those days I thought it to be almost a public duty to oppose in public life. For he was the kind of respectable Liberal who made respectable liberalism stink in the nostrils of so many of my generation who began their political lives as liberals. Winchester and New College, Oxford; Trustee of the British Museum; Vice-Chancellor of Sheffield University; Warden of New College Oxford; he was chosen by Lloyd George, with his unerring instinct for political window-dressing, to be in 1916 Minister for Education (in those days called President of the Board of Education). I may have been prejudiced and unfair, but I thought the Fisher Education Act, which Herbert was responsible for, to be the sort of cowardly compromise which seemed to save the face of its author at the expense of his principles. When he was a Minister, we used to see him fairly often. He would come to us in Richmond, but he also sometimes stayed with Sir Amherst Selby-Bigge, Permanent Secretary at the Board of Education, at Kingston, which was three miles from us at Rodmell, and then he would walk over to see us. His conversation fascinated us; he was so nice, so distinguished, and so ridiculous that he might have walked straight into *Crotchet Castle*. His face and his mind had the gentle, pale, ivory glow, the patina which Oxford culture and innumerable meals at College high-tables give to Oxford dons. So quiet flows the life of the don that there is nearly always something innocent and childlike in his mind. Herbert, with all this academic innocence, suddenly found himself projected into the very centre of the world of action, the House of Commons, Downing Street, the Cabinet. L. G. and the Cabinet went to his head, and he was intoxicated by this 'life of action', though his intoxication, like everything else in him and the Fisher family, was a muted, genteel intoxication. He was obsessed by L. G., who was to him a cross between the superman and a siren, and by Downing Street, sitting in which he felt himself to be sitting bang on the hub of the universe.

About all this poor Herbert discoursed to us lyrically, but with just that touch of humour and restraint required by the good taste which with him was a characteristic both inherited from nineteenth-century ladies and gentlemen and acquired all over again in Oxford. He was never tired of telling us that we and everyone else who did not sit in Downing Street knew nothing about anything. He gave us a vision of the Prime Minister and the President of the Board of Education sitting in the Cabinet Room in Downing Street and receiving an unending stream of secret, momentous messages from every quarter of the earth, if not the remotest galaxies of the outer universe. When the Lloyd George government fell and Herbert went back to New College, he still continued to tell and retell nostalgically the fairy story of his days in Downing Street. And before I leave him in New College I cannot refrain from telling another little absurd story which Adrian Stephen, Herbert's cousin, once told me. Adrian went to stay for a week-end with the

Fishers at New College. The Fisher household was run on extremely economic (to put it euphemistically) lines, which extended to the blankets. It was a bitter cold winter night and Adrian, who was six foot five inches in height, was given a very short bed with a single thin blanket. In the middle of the night he could stand it no longer; he managed to get the whole carpet up and put it over instead of under the bed, and then crept in under it. Unfortunately in the morning he found it impossible to get the carpet properly back in its place and he left his room in a state of chaos. He was not again invited to spend a week-end in New College.

I must return to the election of 1922 and my candidature. I propose to quote from my election address, because it shows where I stood politically a few years after the first great war ended, and also because, I think, my attitude was also that of those who at the time were considered to be on the Left in the Labour Party: we stood between the Labour Party Centre and the Communists and their Fellow Travellers, who were called the Extreme Left. (It has always seemed to me to be curious and confusing that communists are accepted, on their own classification, as Extreme Left; their political outlook and organization is more like that of the Catholic clerical parties on the Continent, the old Centre Party in Germany, and the present Christian Democrats in Italy, and of the deceased Fascists and Nazis, i.e. their correct classification is slightly to the Right of the Extreme Right.) Here then is my declaration of political faith in October 1922:

I am asking for your votes as a candidate adopted by the Seven Universities' Democratic Association. . . . The Association is affiliated to the Labour Party, of which I have been a member for some considerable time. . . . We have in this country two alternatives before us at this election: we can once more entrust the government of the country to one of the two political parties which, for the better part of a century, have separately or in coalition been in power, and which, therefore, are jointly and severally responsible for the social, political, economic, and international conditions in which we find ourselves today; on the other hand, we have an opportunity of making a break with the past and of entrusting the government to a party of new principles and of new men. I confess that one reason why I am a member of the Labour Party, and why ... I ask you to support that Party with your vote, is this: that, looking round upon the political and economic conditions in London and Manchester, in Dublin, India, and Egypt, and remembering the graves in France and Gallipoli which were to be the price of a new world, I feel that this is no time for a mere reshuffle of the ancient Conservative and Liberal Pack and for entrusting power to one or other of the two parties whose political principles and practice are directly responsible for the disastrous situation in which the country finds itself today. A century of Conservative and Liberal Governments brought us war and a peace which has proved hardly better than war. There will be no change if the old men and the old methods are reinstated in Westminster, and if we want a change we must try a party with new principles and new men. I have no illusions with regard to governments and political parties, and I do not ask you to vote for me and for Labour on any promise that we will hang the Kaiser, or make Germany pay, or take five shillings off the income tax, or make everyone peaceful and prosperous. If the Labour Party is returned to power, it will make many mistakes; it will not succeed in carrying out all its principles or all its promises; it will disappoint very many of its supporters. But the Party has this great advantage over the two older

37

parties: unlike them it has ideals and principles which are real and alive, based not upon the possessions and privileges of classes or upon political doctrines which were dead before our grandfathers were born, but upon the generous hopes and vital needs of millions of ordinary men and women. It is these ideals, hopes, and needs which, if I were elected, I should endeavour to help the Labour Party to translate into details of the following practical policy. . . .

There can be no economic recovery in this country, no beginning to build up an educated and prosperous community, unless there is a complete break with the dangerous and extravagant foreign policy which has been pursued equally by Conservative and Liberal Governments. This country must stand out in Europe and the world as a sincere supporter of a policy of peace and international co-operation. The pivot of its programme must therefore be (1) a real League of Nations, inclusive of all nations, the members of which undertake a definite obligation not to go to war; (2) disarmament, beginning with drastic limitation of naval and military armaments, coupled with a general guarantee against aggression; (3) an equitable settlement of the reparation problem and the promotion of good relations between France, Germany, and this country, as the first step towards economic recovery in Europe. This third point is urgent, and it is practicable. The policy which I would support is an offer by this country to France to relieve her of her debt to us in return for her consent (a) to fix reparation payments at a figure which Germany can reasonably be expected to pay, to confine such payments strictly to restoration of the devastated areas, and to grant a sufficiently long moratorium to enable German credit to be restored; (b) to revise the Treaty of Versailles and withdraw the armies of occupation.

I went on to say that I believed 'that the only hope for
Europe consists in the gradual building up of a close understanding and co-operation between Britain, France, and
Germany in a League of Nations', and that the policy out-
lined above was the first step towards such an understanding.
I added four other steps which I considered essential: (1)
recognition of the Russian Government and promotion of
trade with Russia; (2) close co-operation with the U.S.A.;
(3) 'complete abandonment of the policy of imperialism and
economic penetration and exploration which has been pursued by us from time to time in the Near East, Mesopotamia,
Persia, and China'; (4) 'It is essential that the promises of
self-government made to India and Ceylon, and of independence to Egypt, should immediately be carried out with
scrupulous honesty, and, further, that those methods in our
government of the so-called backward races of Africa which
are leading to their subjection and exploitation should be
fundamentally revised'.

In home affairs I said that education was the subject of
greatest importance and I set out an educational policy which
would 'assure to all classes a complete equality of opportunity
to obtain elementary, secondary, and university education'
and which would produce 'an adequate staff of trained and
certified teachers'. I pledged myself to a policy of economy
on unproductive expenditure, a more equitable system of
taxation, and 'a special graduated levy upon fortunes ex-
ceeding £5,000'. I ended my manifesto thus:

In this statement I have confined myself to the immediate and practical problems which will have to be faced
by the next House of Commons. I believe in socialism
and co-operation, but not in violent revolution; I believe
that the resources of the community should be controlled
by and in the interests of the whole community rather than

small groups and classes. But the work of the next Parliament ought to consist neither in bolstering up the vested interests of the present economic system nor in immediately destroying it, but in laying the foundations of a real peace in Europe and of an educated democracy in this country.

I have given at some length this declaration of my political faith in October 1922 both for general, historical, and also for personal reasons. If a man has the temerity to write the story of his life, he should have a double aim: first, to show it and his little ego in relation to the time and place in which he lived his life, to the procession of historical events, even to the absurd metaphysics of the universe; secondly to describe, as simply and clearly as he can, his personal life, his relation, not to history and the universe, but to persons and to himself, his record in the trivial, difficult, fascinating art of living from day to day, hour to hour, minute to minute. I have reached the years in the story of my life which make the first aspect of my autobiography more and more insistent. I have been alive from November 1880 until November 1965; no period in the world's history has been more full of what are called great events, bringing disruption, disaster, cataclysms to the human race, than those 84 years. In the 1914 war there was a nasty poster the object of which was to shame the reluctant citizen into joining the army; it was of a cherubic child asking: 'What did you do in the great war, Daddy?' The question soon lost its meaning—its sting. But 'What did you do in the years between the two great wars?' is a vital question which anyone who took any part in public affairs must answer. The policy for which I declared in the election address was not popular in 1922; it was, and still often is, misrepresented. Looking back over what has happened since, I think that it was the only policy which might have saved Europe from fascism and nazism and from the horror and disaster which they brought upon the world.

I still have to deal with all this later in my passage through the years from 1919 to 1939; now I must return to the election. I rather enjoyed the election campaign, such as it was. The procedure for this peculiar constituency was that candidates made, at most, only one visit to each of the seven universities and made a speech in each. As far as I can remember I visited only Liverpool, Manchester, Durham, and Newcastle. In each I made a speech to what seemed to me a semi-public meeting of already convinced supporters. I do not think that I made a very good impression, partly because I did not always succeed in concealing the fact that I was not really very eager to be an M.P. In March 1921, when I went to Manchester, Virginia came with me. It is many years since I have been in that city, and in 1965 it may be a very different place from what it was when I knew it in the 1920s. In 1921, from the moment when I arrived at its grimy station to the moment when I departed from it, it filled me with a kind of exasperated despair. It was the City of Dreadful Night—'the street lamps burn amidst the baleful glooms'; a drizzle of sooty raindrops dripped remorselessly from the dirty yellow sky upon the blackened buildings and the grey crowds of melancholy men scurrying perpetually, like ants, this way and that way through the foggy streets. Through these streets an unending string of trams ground their way one behind the other; everywhere all the time one's ears were battered by the scraping and grating of their wheels and the striking of their bells. We stayed at the Queen's Hotel, as Virginia recorded in her diary, paying 18s. each for a bed, 'in a large square, but what's a square when the trams meet there? Then there's Queen Victoria like a large tea-cosy, and Wellington, sleek as a mastiff with paw extended.'

41

My spirits, depressed by the streets and sky, by Queen Victoria and the Duke of Wellington, were not raised by my constituents. I made two speeches to them, one before and one after dinner. They were extremely nice and extremely good people, many of them professors or lecturers, who had been conscientious objectors or had been arrested for keeping the flag of liberty flying in Manchester during the war. But, like Queen Victoria and the Duke, they were somewhat grey, depressed, low in tone. 'Old Mrs Hereford and Professor Findlay', Virginia noted, 'sat patiently looking at the tablecloth with nothing to say, like two old horses who have been working in the fields all day together.' When Mrs Findlay asked Virginia whether she was a politician and whether she did much organizing work, Virginia said she listened. 'Mrs Findlay shook her head. Why was I there then?' There was, it must be admitted, no satisfactory answer to that question.

When I find myself in a strange city, at a loose end, waiting as one does eternally in strange cities for a boat, a plane, or an interview—when time seems to stop and the universe seems to have dwindled to an unending series of hotel corridors, lavatories, and lounges—I tend to go to the zoo. I am ambivalent about zoos: I have an uneasy feeling that one should not keep animals in cages, but I never get tired of watching animals anywhere. You can learn a great deal about the character of a country or city by going to its zoo and studying its arrangement and the behaviour of the animals. The London Zoo is an animal microcosm of London, and even the lions, as a rule, behave as if they had been born in South Kensington. I once saw a curious incident there when one of them did not. It was a warm summer day and the lions were in their outdoor cages. A stout, middle-aged, middle-class lady was standing near the bars looking at a magnificent lion who was standing on the other side of the bars gazing over her head, as lions seem to do, into eternity.

Suddenly he turned round, presented his backside to her, and pissed on her through the bars. 'Oh the dirty beast,' she said, 'Oh, the dirty beast,' wiping her face and blouse, half angry, half amused, and the tone of her voice was exactly as if some nasty little boy had done some dirty trick in the Earls Court Road.

I recall vividly two other zoos. When I was in Jerusalem in 1957, I had to go to the Foreign Office which seemed to me a long weary way out from the centre of the city. I started to walk back; it was hot and dusty, and I seemed to have got into a ramshackle suburb frequented by those unshaven, long-haired orthodox Jews, young men whose self-conscious, self-righteous hair and orthodoxy fill me with despair. When I saw a signpost directing me to the zoo, I made off for it at once. But I did not, as I had hoped, escape from the melancholy of the dreary streets and the moth-eaten anachronism of those ridiculous young men. I have never seen anywhere else so melancholy a collection of animals. The architecture of the zoo seemed to be a ramshackle replica of the surrounding streets, and long-haired monkeys gazed at one, it seemed to me, with the self-satisfaction of all the orthodox who have learned eternal truth from the primeval monkey, all the scribes and pharisees who spend their lives making mountains of pernicious stupidity out of molehills of nonsense.

I should add that the long-haired monkeys in the Jerusalem Zoo and the long-haired orthodox Israelis in the neighbouring streets are characteristic of only one side of contemporary Israel. When you enter Israel by Tel Aviv, buzzing with business as though it were a gigantic human hive, drive up the road to Jerusalem, or visit Haifa and Tiberias, you are exhilarated by the energetic exhilaration of the people who are everywhere living dangerously and happily wresting from the rocky earth and a ring of implacable enemies a new way of life in a strange land. These people, who form the immense majority of the population, are the exact opposite of the orthodox Jews in Jerusalem. It makes it the more lamentable that they should allow politics and therefore life in Israel to be continually influenced by orthodox Judaism. Again and again one has in life to say, with Lucretius, '*tantum religio potuit suadere malorum*'—how much evil religion has induced human beings to do! One is accustomed to see throughout history down to today the absurd delusions of savages promoted to divine truths and their morality and rules of conduct maintained for two or three thousand years as an excuse for protecting the vested interests of ignorance and injustice. To see this process once more repeated in modern Israel is horrifying. After all, the austere, bare monotheism which the ancient Hebrews developed made it comparatively easy for their Jewish descendants in modern times to shed the primitive beliefs and rituals and morality of the Pentateuch. Already two thousand years ago the writers of the books which we call *Job*, *Ecclesiastes*, and *Micah* had laid the foundations of a civilized morality and a sceptical, rational theism, from which by the process of time might come 'the religion of all sensible men'—agnosticism or atheism. It is deplorable to find the builders of the modern state in Israel making its laws conform to the belief that the creator of the universe, with its suns, planets, galaxies, atoms—old bearded Jehovah sitting up there on Mount Sinai amid the thunder and lightning and once in three thousand years showing his backside to a favoured Moses—that this omnipotent deity has enacted an eternal law, revealed to a handful of rabbis and ignorant men in Jerusalem, regulating the shaving and haircutting of males, the eating of pork and the slaughtering of sheep, and the use of trains and taxis on days which some people happen to call Saturday and others Samedi.

The other zoo which I vividly recall is in Colombo. Colombo is as different from Jerusalem as the Sinhalese are from the Israelis. My likes and dislikes are catholic, and I have remarked before in this autobiography that I see no reason why, because one likes claret, one should not also like burgundy; I like both, and I like both Israel, with its fierce sun, fiery rocks, its furnace of human activity, and Ceylon, with its tropical 'lilies and languors', the gaily coloured kaleidoscope of flowers, trees, and cheerful, drifting crowds. Three years after I visited the Jerusalem Zoo I visited Ceylon. The plane which was to take me back to England was delayed somewhere in eastern Asia and I found myself in one of those exasperating predicaments in which one has nothing to do but wait indefinitely for someone to ring you up and say that you need wait no longer. I waited from 8 in the morning, when my plane should have left, until 12 midnight, when it did. Half-way through the morning I went to the Colombo Zoo. It was a microcosm of Colombo, of the Sinhalese low country, of the Sinhalese way of life. It was 'full of trees and waving leaves', amid which, in the humid languorous heat, elephants, lions, leopards, bears lived their happy natural lives. I was watching a family of lions, father, mother, and three cubs, who were in a large open-air enclosure. The male was lying asleep up in one corner and the female was drowsing down in the other corner; the cubs were playing about. Suddenly one of the cubs went over and began to play with his father's tail. There was a low growl; the tail flapped angrily on the ground; the cub made another dart at it. The lion lifted his great head just off the ground and let out a blood-curdling snarl. The lioness rose up and slowly, threateningly went over to the lion and quite silently stood over him between him and the cub. The lion got up and slowly, sullenly walked over to the farthest corner where he flopped down

45

to sleep again. The lioness shooed the cub back to her corner. It seemed to me that I might have been watching a domestic scene in a compound in one of those Sinhalese villages, a Sinhalese mother slowly, firmly shooing her child away from mischief and danger.

I must unfortunately return once more from the Colombo Zoo in 1960 to Manchester, Durham, and the election in 1922. When I went to meet my constituents and make a speech to them in Durham, the meeting was held in the room of a graduate. The audience was extremely small; indeed, in each of the universities the number of people who came to my meetings was very small. I was told by the chairman of the Durham meeting that many who would support and might vote Labour would not do anything openly which might connect them with the Labour Party, because that connection would be viewed with disapproval by the university authorities and would jeopardize their prospects of a good job after they had taken a degree. He said that this was to some extent due to political prejudice, but what was more important was the fact that the university relied for financial and other support in part upon the local wealthy Conservatives, often 'big business', and had to be careful not to offend or antagonize them. Theoretically and on the surface your religion, politics, or economics were as unimportant as the colour of your hair, but if you wanted to do well for yourself academically it was safer to conceal the fact that you were left of centre. I was told that this was more or less true of all the provincial universities in those bad old days.

I came, as was expected, at the bottom of the poll, Sir Martin Conway, Conservative, and the Right Honourable Herbert Fisher, Liberal, being elected. Conway was a curious man. He had been mountaineer and explorer and a Slade Professor, and in 1922 he was Director General of the Imperial War Museum. Shortly after the election I received the following letter from him:

It was only two days ago that I learned who you are—the author of *A Village in the Jungle*. That is a book which I read with extraordinary delight and which I treasure alongside of the *Soul of a People*. I am really sorry to have been put in opposition to a writer I so heartily admire. Of one thing, however, I am certain: the writer of such a book would have found the H. of C. a most unattractive place and would have been very unhappy there, especially if he had been obliged to associate intimately with the rank and file of the Labour Party—tho', of course, among them are some delightful simple souls, very lovable, but the bulk are not such.

I am venturing to send you a book of mine in kindly remembrance of our contest. Your late father-in-law was an honoured friend of mine, as you may guess.

Yours faithfully
Martin Conway

Sir Martin asked me to lunch with him in the House of Commons and I found him to be a pleasant, not very interesting man. His letter—and the attitude towards the Labour Party in the Universities—show clearly the social and political snobbery of those days. It amused me that the Slade Professor of Fine Arts should think that I must be such a sensitive plant that I would wilt unhappily in the company of the rough trade unionists of the Labour Party. As a matter of fact I knew a good many Labour M.P.s as Secretary of the Labour Party Advisory Committees and because for a time I was Parliamentary Correspondent of the *Labour Leader*. Of course, some of them were pretty tough, but I never found any difficulty in getting on with them.

In the period of my life of which I am now trying to tell

the story—and indeed in the whole of my life after the year 1919—it has, as I have said, been dominated by politics and public events which are living and lived history. As I look back over the mental and physical chaos and kaleidoscope which has been my individual life from 1919 to 1965, I see that history ruthlessly divided it into four periods: (1) 1919 to 1933, the fourteen years of struggle for civilization which ended with Hitler's rise to power; (2) 1933 to 1939, the six years in which civilization was finally destroyed and which ended with war; (3) 1939 to 1945, the six years of war; (4) the post-war world. If I am to continue with the story of my life, I shall have to deal with the events of each of these four periods, the effect of each upon me and my life, and my reaction to each of them. But even in our cruel, mechanized, barbarous age, we have not yet become completely robots, puppets jerked through life by history, governments, and computers. We still have, at any rate in Britain, some shreds of private life, which we can preserve unaffected by public events. It is to our private lives that I must now turn—to return later to politics in the kind of see-saw which must inevitably continue through the fourth volume of my autobiography. What dominates or moulds our private lives privately is, as I have said, the house in which we live. In each period of our living we are profoundly influenced, therefore, by both history and geography, by time and place. In my own case, war and peace and Stalin, Mussolini, and Hitler divided the twenty years of my life from 1919 to 1939 into the two periods 1919-1933 and 1933-1939, and the same twenty years were deeply divided for us into two other periods by two houses, Hogarth House in Richmond from 1919 to 1924 and Tavistock Square in Bloomsbury from 1924 to 1939. I must now deal with the period of six years in Richmond.

One must begin with Virginia's illness and her slow

recovery from it, for we continued to live in Richmond mainly to protect her from London and the devastating disorientation which would threaten her from social life if we returned to live there. It was a perpetual struggle to find the precarious balance of health for her among the strains and stresses of writing and society. The routine of everyday life had to be regular and rather rigid. Everything had to be rationed, from work and walking to people and parties. Despite all our precautions, her diary shows how often in the first few years after the war she was ill or threatened with illness. The threat was almost always a headache, which was the warning signal of mental strain; the 'illness', if it came to that, was the first stage towards breakdown. We knew exactly what the treatment should be; the moment the headache came, she had to go to bed, and remain there comatose, eating and sleeping, until the symptoms began to abate. That was the cure; the difficulty was always to perform the actions which the cure required: unfortunately one does not sleep—or even eat—because one knows that sleeping or eating is the one thing which will cure one of a shadow across one's brain.

In 1921 and 1922 Virginia was continually beset with these attacks. For instance, in the 1921 diary there is an entry for June 7 describing how Tom Eliot came to tea and joined us in lamenting the Pecksniffian character of John Middleton Murry. The next entry in the diary is August 8 and I will quote it, because it shows so clearly what Virginia actually suffered in one of these threatening attacks:

What a gap! How it would have astounded me to be told when I wrote the last word here, on June 7th, that within a week I should be in bed, and not entirely out of it till the 6th of August—two whole months rubbed out— These, this morning, the first words I have written—to call writing—for 60 days; and those days spent in wearisome headache, jumping pulse, aching back, frets, fidgets, lying awake, sleeping draughts, sedatives, digitalis, going for a little walk, and plunging back into bed again—all the horrors of the dark cupboard of illness once more displayed for my diversion. Let me make a vow that this shall never, never, happen again; and *then* confess that there are some compensations. To be tired and authorized to lie in bed is pleasant; then scribbling 365 days of the year as I do, merely to receive without agitation of my right hand in giving out is salutary. I feel that I can take stock of things in a leisurely way. Then the dark underworld has its fascinations as well as its terrors; and then sometimes I compare the fundamental security of my life in all (here Mrs Dedman interrupts for 15 minutes) storms (perhaps I meant) with its old fearfully random condition. Later I had my visitors, one every day, so that I saw more people than normally even. Perhaps, in future I shall adopt this method more than I have done. Roger, Lytton, Nessa, Duncan, Dorothy Bussy, Pippa, Carrington, James and Alix—all these came; and were as detached portraits—cut out, emphatic, seen thus separately compared with the usual way of seeing them in crowds. Lytton, I note, is more than ever affectionate. One must be, I think, if one is famous. One must say to one's old friends 'Ah my celebrity is nothing—nothing—compared with this'.

This was, of course, a severe bout, and in the autumn and winter of 1921 she had recovered to her normal equilibrium which allowed her safely—but within limits—to work and live a social life. But during the first seven months of 1922 she was off and on continually unwell or threatened with headaches. In March she started a temperature which the doctors took seriously and sent us on a fairly long odyssey through Harley Street and Wimpole Street which gave us a curious view of medical science and the tiptop Harley Street specialists. We had at the time an extremely nice, sensible G.P., Dr Ferguson. He sent us first to a lung specialist who said that Virginia's symptoms were due to her lungs, which were in a serious state. When this was reported to Ferguson, he said it was nonsense; he had examined her lungs frequently and there was nothing wrong with them; we should ignore the diagnosis. He sent us off to a heart specialist, who said that Virginia's symptoms were due to her heart, which was in a serious state. We returned sadly to Richmond and Ferguson. I was told by him that the great man had diagnosed inflammation of the heart, a disease from which some famous man—I think it was the great Alfred Harmsworth, Lord Northcliffe—had just died. The disease was incurable and death imminent and inevitable. In his opinion, he said, this was nonsense; he had frequently examined her heart and there was nothing seriously wrong with it; we should ignore the diagnosis. We did so. We went, I think, to one more specialist, a distinguished pathologist who discovered that Virginia was suffering from the disease in which he specialized. He was wrong; we ignored his diagnosis and decided to forget about it and about Harley Street. She not only recovered from the three fatal and incurable diseases; the disquieting symptoms gradually disappeared.

At our last interview with the last famous Harley Street specialist to whom we paid our three guineas, the great Dr Saintsbury, as he shook Virginia's hand, said to her: 'Equanimity—equanimity—practise equanimity, Mrs Woolf'. It was, no doubt, excellent advice and worth the three guineas, but, as the door closed behind us, I felt that he might just as usefully have said: 'A normal temperature—ninety-eight point four—practise a normal temperature, Mrs Woolf'.

With regard to her writing, Virginia certainly never learned to practise equanimity. Like most professional writers, if she was well, she went into her room and sat down to write her novel with the daily regularity of a stock-broker who commutes every day between his house in the suburbs and his office in the neighbourhood of Throgmorton Street. Her room was very different from a stock-broker's office. She was an untidy writer, indeed an untidy liver, an accumulator of what Lytton Strachey used to call 'filth packets', those packets of old nibs, bits of string, used matches, rusty paper-clips, crumpled envelopes, broken cigarette-holders, etc., which accumulate malignantly on some people's tables and mantelpieces. In Virginia's workroom there was always a very large, solid, plain wooden table covered with filth packets, papers, letters, manuscripts, and large bottles of ink. She very rarely sat at this table, certainly never when she was writing a novel in the morning. To write her novel of a morning she sat in a very low armchair, which always appeared to be suffering from prolapsus uteri; on her knees was a large board made of plywood which had an inkstand glued to it, and on the board was a large quarto notebook of plain paper which she had bound up for her and covered herself in (usually) some gaily-coloured paper. The first draft of all her novels was written in one of these notebooks with pen and ink in the mornings. Later in the morning or in the afternoon, or sometimes at the beginning of the next morning, she typed out what she had written in the notebook, revising it as she typed, and all subsequent revisions were made on the typewriter. A curious thing about her was that, although she was extremely sensitive to noise and was one of those people who 'jumped out of her skin' at a sudden noise or unexpected confrontation, she seemed usually, when writing, to acquire a protective skin or integument which insulated her from her surroundings. Her room tended to become not merely untidy but squalid. She reached the final stage of organized disorganization and discomfort when we moved from Richmond to 52 Tavistock Square in Bloomsbury. At the back of the house was what had once been an immense billiard room. We used it as a storeroom for the Hogarth Press and there embedded among the pyramids and mountains of parcels, books, and brown paper sat Virginia with her disembowelled chair, her table, and her gas fire.

In the regularity of this routine of writing and in her disregard of her surroundings when writing one might not unreasonably have seen a measure of equanimity. Up to a point this was true; in some ways her attitude in writing and to her writing was extraordinarily controlled, dispassionate, coldly critical. In the process of her writing—of her artistic creation—there were long periods of, first, quiet and intense dreamlike rumination when she drifted through London streets or walked across the Sussex water-meadows or merely sat silent by the fire, and secondly of intense, analytical, critical revision of what she had written. No writer could possibly have given more time and intensive thought to the preparation for writing and to the revision of what she had written. Both these periods required and got from her dispassionate equanimity. But there were also for her two periods of passion and excitement. The first was in the moment of creation, in the whole process of actual writing. I think that, when writing, Virginia was almost the whole time writing with concentrated passion. The long strenuous intellectual process was over and would be called in again for revision; now emotion and imagination took control. And at moments, as I pointed out in *Beginning Again*,[1] genius or inspiration seemed to take control, and then, as she described how she wrote the last pages of *The Waves*, 'having

[1] Page 31.

reeled across the last ten pages with some moments of such intensity and intoxication that I seemed to stumble after my own voice, or almost after some sort of speaker (as when I was mad) I was almost afraid, remembering the voices that used to fly ahead'.[1] There was, of course, no place or possibility for Dr Saintsbury's 'equanimity' in this kind of emotional and imaginative volcanic eruption, the moment of artistic creation. But I think that whenever Virginia was actually writing a novel—or rather the first draft of a novel—her psychological state was in a modified degree that described above. The tension was great and unremitting; it was emotionally volcanic; the conscious mind, though intent, seemed to follow a hair's breadth behind the voice, or the 'thought', which flew ahead.[2] It was this terrific, persistent tension which, because it naturally produced mental exhaustion, made her writing a perpetual menace to her mental stability. And the moment that the symptoms of mental exhaustion began, she was unable to write.

In what I have written above, I have distinguished two markedly different—indeed almost antithetical—phases in Virginia's creative process. This swing of the pendulum in the mind between conscious, rational, analytic, controlled thought and an undirected intuitive or emotional process almost always takes place where the mind produces something original or creative. It happens with creative thinkers, scientists, or philosophers, no less than with artists. Perhaps the most famous instance was recorded over two thousand

[1] *A Writer's Diary*, p. 165.
[2] I think that when she was revising and rewriting on a typewriter what she had written with pen and ink in the morning, her psychological state and her method were quite different. The conscious, critical intellect was in control and the tension was less. It was largely the same when she was writing criticism. I used to say that, when she came in to lunch after a morning's work, I could tell by the depth of the flush on her face whether she had been writing fiction or criticism.

years ago in Sicily when the problem in hydrostatics which Archimedes had unsuccessfully worked upon for days suddenly solved itself in his drowsing mind as he lay in his bath and he dashed out naked into the streets of Syracuse shouting: 'I have found it, I have found it!' Virginia, too, often 'found it' by the same kind of mental process and with the same excitement.

It was this excitement which was the sign and symptom of the mental strain of her writing and which was a perpetual menace to her stability. But there was also, as I have said above, a second period of passion and excitement through which she almost always had to pass in the process of writing a novel. This came upon her almost invariably as soon as she had finished writing a book and the moment arrived for it to be sent to the printer. It was a kind of passion of despair, and it was emotionally so violent and exhausting that each time she became ill with the symptoms threatening a breakdown. In fact, the mental breakdown which I described in *Beginning Again* occurred immediately after she had finished writing *The Voyage Out*, and the breakdown in 1941 which ended in her suicide occurred immediately after she had finished *Between the Acts*. And in 1936, when she had finished *The Years* and she had to begin correcting the proofs, she came desperately near a mental breakdown. On April 19 she wrote in her diary:

> The horror is that tomorrow, after this one windy day of respite—oh the cold north wind that has blown ravaging daily since we came, but I've had no ears, eyes, or nose: only making my quick transits from house to room, often in despair—after this one day's respite, I say, I must begin at the beginning and go through 600 pages of cold proof. Why, oh why? Never again, never again.[1]

[1] *A Writer's Diary*, p. 259.

The psychology of the artist in the final stages of creation and production is very interesting. Many writers have I think felt, but have not, like Virginia, recorded, the horror of facing those pages of 'cold proof' and, even more, the cold breath of criticism in the first days after publication. In *Beginning Again* I said that, in my opinion, one reason why Desmond MacCarthy never wrote the novel which, when he was a young man, we thought he would write and which he intended to write, was that he could not face the responsibility of publication, the horror of the final day when the book and the author are handed over to the icy judgment of the reviewers and the public. However sensitive the serious author may be, the moment comes when he has to be ruthless with himself. He must coldly go through the weary waste of cold proof, put the last comma into the last sentence, deliver himself artistically naked to the public, take the icy plunge of publication. At that point he must have the courage to say to himself: 'Literary editors, reviewers, my friends, the great public—they can say what they like about the book and about me. Hippokleides doesn't care.'[1] Virginia was terribly —even morbidly—sensitive to criticism of any kind and from anyone. Her writing was to her the most serious thing in life, and, as with so many serious writers, her books were

[1] I have the greatest admiration for Hippokleides—his story is in Herodotus. He became engaged to Kleisthenes's daughter, and at the feast to celebrate the engagement he got rather above himself and danced on his head on the table. When Kleisthenes, who was very grand, being Dictator of Sikuon, saw the legs waving in the air, he was outraged and said: 'O son of Tisandros, you have danced away your marriage'. To which Hippokleides replied: 'Hippokleides doesn't care'. His reply became a Greek wisecrack. How surprised Kleisthenes and, indeed, Hippokleides would be at the vagaries of immortality if they knew that the Dictator and hundreds of other 'great' men of his time are completely forgotten while Hippokleides, with his legs waving in the air above his head and his wisecrack, are still, 2,400 years later, recalled with approval and affection.

to her part of herself and felt to be part of herself somewhat in the same way as a mother often seems all her life to feel that her child remains still part of herself. And just as the mother feels acutely the slightest criticism of her child, so any criticism of her book even by the most negligible nitwit gave Virginia acute pain. It is therefore hardly an exaggeration to say that the publication of a book meant something very like torture to her.

The torture began as soon as she had written the last word of the first draft of her book; it continued off and on until the last reviewer, critic, friend, or acquaintance had said his say. And yet, despite her terrifying hypersensitivity, there was in Virginia an intellectual and spiritual toughness which Desmond lacked. It came out in the dogged persistence with which she worked at every word, sentence, paragraph of everything she wrote, from a major novel to a trumpery— or what to almost any other writer would have seemed a trumpery—review. The consequence was that the moment always came when she stiffened against the critics, against herself, and against the world; ultimately she had the courage of her convictions and published, saying—not with much conviction—'Virginia doesn't care'. And that was why, unlike Desmond, she had published, when she died, seventeen books.

The psychology of human misery is curious and complicated. Several critics have expressed surprise and disapproval at the spectacle of Virginia's misery over blame or even the lack of praise. Max Beerbohm, for instance, 'had reservations' about her and disliked her diary. 'I have never understood,' he said, 'why people write diaries. I never had the slightest desire to do so—one has to be so very self-conscious.' (It is significant that Max thought that his not wanting to do something was a good reason or an excuse for his not understanding why other people wanted to do it and did it.) 'It was deplorable,' he said, 'to mind hostile criticism as much as Virginia Woolf did.'[1] It was no doubt highly deplorable both ethically and from the point of view of her own happiness. But it and the writing of diaries is surely a little less difficult to understand than Max found them. The mother's instinct to resent criticism of her children is irrational and deplorable, but common and not entirely unnatural. Vanity explains part of it, but not, I think, all. Oddly enough there is mixed in with the vanity something which is almost the opposite of vanity in these cases, a kind of objective ideal. The mother wants the child to be perfect for its own sake, and Virginia, whose attitude towards her books was, as with so many serious writers, maternal, wanted her books to be perfect for their own sake. She was also abnormally sensitive, both physically and mentally, and it was this which helped her both to produce her novels and to be miserable when Mrs Jones or Max Beerbohm didn't like them. And I rather think it helped her once to hear the sparrows talking Greek outside her bedroom window.

When I remember how, owing to her health, Virginia always had to restrict her daily writing to a few hours and often had to give up writing for weeks or even months, how slowly she wrote and how persistently she revised and worked over what she had written before she published it, I am amazed that she had written and published seventeen books before she died. There are today (in 1965) 21 volumes in the list of her publications.[2] This is the more surprising because none of these books had been written by her before the age of 30, and all her major works, except *The Voyage Out* and *Night and Day*, were written in the last 21 years of her life. In the period with which I am now concerned, our years at Richmond from 1919 to 1924, her writing was strictly rationed and often interrupted. They were years of

[1] See *Max*, by Lord David Cecil, pp. 483-484. [2] In England.

crucial importance in her development as a novelist, for during them she revolted against the methods and form of contemporary fiction—pre-eminently the fiction of Galsworthy, Wells, and Bennett—and created the first versions of her own form and methods which ultimately and logically developed into those of *The Waves* and *Between the Acts*.

The process began with the 'short stories' which she collected and published in March 1921 in *Monday or Tuesday*. They were all written between the years 1917 and 1921 and were a kind of prelude or preliminary canter to *Jacob's Room* which was published in October 1922. *The Mark on the Wall* was the first sign of the mutation in method which was leading to *Jacob's Room*; it was published in July 1917, forming part of *Two Stories*, our first Hogarth Press publication, and was written at the end of 1916 or the beginning of 1917. It has been said by some critics that Virginia derived her method, which they call the stream of consciousness, from Joyce and Dorothy Richardson. The idea that no one in the arts has ever invented anything or indeed has ever had an original thought, since everything is always 'derived' from something else in an unending artistic House that Jack Built, is extremely common and has always seemed to me untrue—and if not untrue, unimportant. The merits or defects of *The Waves* remain unaffected whether they were or were not 'influenced' by Joyce's *Ulysses* or Dorothy Richardson's *The Tunnel*. But it is perhaps just worth while to point out that *The Mark on the Wall* had been written at latest in the first part of 1917, while it was not until April 1918 that Virginia read *Ulysses* in manuscript and January 1919 that she read *The Tunnel*.

In May 1919 we published in the Hogarth Press *Kew Gardens*. As I recorded in *Beginning Again* (p. 241), this thin little volume, which we had printed ourselves, had great importance for us, for its immediate success was the first of many unforeseen happenings which led us, unintentionally and often reluctantly, to turn the Hogarth Press into a commercial publishing business. But it was also a decisive step in Virginia's development as a writer. It is in its own small way and within its own limits perfect; in its rhythms, movement, imagery, method, it could have been written by no one but Virginia. It is a microcosm of all her then unwritten novels, from *Jacob's Room* to *Between the Acts*; for instance, Simon's silent soliloquy is a characteristic produced by the same artistic gene or chromosome which was to produce 12 years later Bernard's soliloquy in *The Waves* and 22 years later the silent murmurings of Isa in *Between the Acts*.

Virginia began to write *Jacob's Room* in April 1920; the period of eleven months between that date and May 1919, when *Kew Gardens* was published, had been one of rumination and preparation. She wrote little, and what she did write was journalism. It was for her a disturbed period, partly because of warning headaches and partly because of the publication of *Night and Day* in October. It is clear from the diary that 'the creative power' which, Virginia said, 'bubbles so pleasantly in beginning a new book' was simmering throughout those eleven months. By April 1920 the content, characters, form of *Jacob's Room* must have been in her mind in some detail, for she had already given the book its significant name. And she was conscious that the method of her experimental short stories, *The Mark on the Wall*, *Kew Gardens*, and *An Unwritten Novel*, must be adapted to produce a full-length novel. This is what she wrote in her diary on January 26, 1920 (*A Writer's Diary*, p. 22):

Suppose one thing should open out of another—as in an unwritten novel—only not for 10 pages but 200 or so —doesn't that give the looseness and lightness I want; doesn't that get closer and yet keep form and speed, and enclose everything, everything? My doubt is how far it will enclose the human heart—Am I sufficiently mistress of my dialogue to net it there? For I figure that the approach will be entirely different this time: no scaffolding; scarcely a brick to be seen; all crepuscular, but the heart, passion, humour, everything, as bright as fire in the mist. Then I'll find room for so much—a gaiety—an inconsequence— a light spirited stepping at my sweet will. Whether I'm sufficiently mistress of things—that's the doubt; but conceive *Mark on the Wall*, *K. G.*, and *Unwritten Novel* taking hands and dancing in unity.

On November 6, 1921, she wrote the last words of *Jacob's Room*; it was published in October 1922. The writing of this novel was the beginning of a period of great fertility. In the three years 1921 to 1924 she wrote *Mr Bennett and Mrs Brown* (published in 1924), and prepared or wrote the material included in *Monday or Tuesday* (published in 1921) and *The Common Reader* (published in 1925). In 1922 she began *Mrs Dalloway* and was writing it all through 1923 and 1924. She began it as a short story and for a short time hesitated whether to expand it into a full-length novel. Early in 1923 she had decided upon its being a novel; she called it at first *The Hours*, but finally went back to the original title *Mrs Dalloway*.

Throughout these six years, 1919 to 1924, she was also writing a considerable amount of journalism, if one remembers how much work she was giving to her books and how limited was the total time she could devote to writing. Most of her journalism consisted of reviews in the *Times Literary Supplement* and, after 1923, in the *Nation*. Her attitude to her reviewing was not consistent. She looked upon it usually as a method of making money—at this time, indeed, it was almost her only way. As such she resented it and occasionally decided to give it up altogether. The following entry in her diary on September 15, 1920, shows this:

> I should have made more of my release from reviewing. When I sent my letter to Richmond,[1] I felt like someone turned out into the open air. Now I've written another in the same sense to Murry,[2] returning Mallock; and I believe this is the last book any editor will ever send me. To have broken free at the age of 38 seems a great piece of good fortune—coming at the nick of time, and due of course to L., without whose journalism I couldn't quit mine. But I quiet my conscience with the belief that a foreign article once a week is of greater worth, less labour and better paid than my work; and with luck, if I can get my books done, we shall profit in moneymaking eventually. And, when one faces it, the book public is more of an ordeal than the newspaper public, so that I'm not shirking responsibility. Now, of course, I can scarcely believe that I ever wrote reviews weekly; and literary papers have lost all interest for me. Thank God, I've stepped clear of that *Athenaeum* world, with its reviews, editions, lunches, and tittle tattle—I should like never to meet a writer again. The proximity of Mr Allison,[3] reputed editor of the *Field*, is enough for me. I should like to know masses of sensitive, imaginative, unselfconscious, unliterary people, who have never read a book. Now, in the rain, up to Dean, to talk about the door of the coal cellar.

But Virginia did not always hold this uncompromising view of her journalism and reviewing. She found that she

1 Bruce Richmond, editor of the *Times Literary Supplement*.
2 J. Middleton Murry, editor of the *Athenaeum*.
3 Wrongly reputed. He was advertisement manager of *The Times* and owner of the *London Mercury*. He had just bought a large house and farm in Rodmell.

could not go on for long periods uninterruptedly writing fiction and she relieved the strain by doing something which used another part of her brain or literary imagination. As the years went on, she discovered that reviewing performed this function admirably; it gave her the relief which some thinkers or writers find in chess or crossword puzzles.

I give below the figures of Virginia's earnings by her pen or typewriter in the years 1919 to 1924. They are interesting, I think, from a particular point of view, from the light which they throw on her work as a novelist and journalist, but also from a general point of view, the economics of the literary profession in the 1920s.

	Journalism			Books			Total		
1919	£153	17	0	nil			£153	17	0
1920	234	6	10	£106	5	10	340	12	8
1921	47	15	1	10	10	8	58	5	9
1922	69	5	0	33	13	0	102	18	0
1923	158	3	9	40	0	5	198	4	2
1924	128	0	0	37	0	0	165	0	0

Virginia was 42 years old in 1924. Jane Austen died at the age of 42, Emily Brontë at the age of 30, Charlotte Brontë at the age of 39; all three were famous novelists and had written best-sellers at the time of their deaths. By the age of 42 Virginia had already published three major novels: *The Voyage Out*, *Night and Day*, and *Jacob's Room*. All three had been widely recognized as novels of great merit, and even genius; they had been published in America as well as in Britain; but their sales were small in both countries. Thus the total which she earned from her books, including the American editions, in the six years, ending with 1924 was £228, or £38 per annum; indeed her total earnings, from books and journalism, during the period were only £1,019, or £170 per annum.

Many people, including even many writers and publishers, will be surprised at these miserable figures. Having for 50 years observed, as writer, editor, and publisher, the rise and fall of many literary reputations and incomes, I know that there is nothing particularly unusual in them. Nothing can be more erratic and fickle than literary reputations and earnings. In 1963, i.e. 40 years after 1924, these three novels earned in royalties in Britain alone £251—£22 more than all Virginia's books, including the three novels, had earned in Britain and America during the six years. There must be many best-sellers of the year 1924 which did not sell a copy or earn their authors a penny in the year 1963. I shall later on give from time to time the figures of Virginia's sales and earnings; here I will give only one other fact to show how sudden and unpredictable are the movements in the market where writers sell their wares. Five years after 1924, the year in which Virginia earned £38 by her books, she earned £2,063 by her books. This astronomical increase after the six years of complete stagnation was due partly to *Mrs Dalloway*, published in 1925, and *To the Lighthouse*, published in 1927, but still more to *Orlando*, which was published in 1928.

The development of the Hogarth Press was bound up with the development of Virginia as a writer and with her literary or creative psychology. When we moved from Hogarth House, Richmond, to 52 Tavistock Square on March 13, 1924, the Hogarth Press had published 32 books in the seven years of its existence. I give below the complete list of books published in each of these seven years, for it shows the scale and quality of the development. I have marked with an asterisk the books which we printed with our own hands.

1917 *Two Stories* by Leonard and Virginia Woolf
1918 *Prelude* by Katherine Mansfield

1919   *Poems* by T. S. Eliot
       Kew Gardens by Virginia Woolf
       Critic in Judgment by J. Middleton Murry

1920   *Story of the Siren* by E. M. Forster
       Paris by Hope Mirrlees
       Gorky's *Reminiscences of Tolstoi*
       Stories from the Old Testament by Logan Pearsall Smith

1921   *Monday or Tuesday* by Virginia Woolf
       Stories from the East by Leonard Woolf
       Poems by Clive Bell
       Tchekhov's Notebooks

1922   *Jacob's Room* by Virginia Woolf
       Stavrogin's Confession by Dostoevsky
       The Gentleman from San Francisco by Bunin
       Autobiography of Countess Tolstoi
       Daybreak by Fredegond Shove
       Karn by Ruth Manning-Sanders

1923   *Pharos and Pharillon* by E. M. Forster
       Woodcuts by Roger Fry
       Sampler of Castile by Roger Fry
       The Waste Land by T. S. Eliot
       The Feather Bed by Robert Graves
       Mutations of the Phoenix by Herbert Read
       Legend of Monte della Sibilla by Clive Bell
       Poems by Ena Limebeer
       Tolstoi's Love Letters
       Talks with Tolstoi by A. V. Goldenveiser
       Letters of Stephen Reynolds
       The Dark by Leonid Andreev
       When it was June by Mrs Lowther

The above list shows that, though we started the Hogarth Press in 1917, it was only in 1920 that we began—with Gorky's *Reminiscences of Tolstoi* and Logan Pearsall Smith's

Stories from the Old Testament—to have books printed for us
by commercial printers and so to become ourselves commercial publishers. The four books published by us in 1917,
1918, and 1919 were all printed and bound by ourselves,[1]
and the production, such as it was, was entirely that of the
hands of Leonard and Virginia Woolf. Until 1920 the idea
of seriously becoming professional publishers never occurred
to us. The Hogarth Press was a hobby, and the hobby consisted in the printing which we did in our spare time in the
afternoons. A second object, which developed from the
first, was to produce and publish short works which commercial publishers could not or would not publish, like
T. S. Eliot's poems, Virginia's *Kew Gardens*, and Katherine
Mansfield's *Prelude*. We were able to do this without financial loss, because we printed and bound the books ourselves
in the dining-room or basement of Hogarth House and had
no 'overheads'. Our first step up or down into professional
publishing was the result of the sudden success of *Kew
Gardens*. As I explained in *Beginning Again* (p. 241), when
we were suddenly overwhelmed with orders for the book
from booksellers, we decided to have a second edition printed
for us commercially. This brought us into contact with all the
big and many of the small booksellers, both wholesale and
retail, and it was not difficult to learn the not very complicated customs and structure of the book trade. *Kew Gardens*
showed me that we could, if we wished, publish a book
commercially and successfully.

When, therefore, Koteliansky brought us Gorky's

[1] On the title-page of our first book, *Two Stories*, we put 'Written
and printed by Virginia Woolf and L. S. Woolf' and the imprint was
Hogarth Press, Richmond. Later on our usual imprint on books
printed by ourselves was 'Printed and published by Leonard and
Virginia Woolf at The Hogarth Press' and on books printed for us
by a commercial printer 'published by Leonard and Virginia Woolf
at The Hogarth Press'.

Reminiscences of Tolstoi and suggested that we should publish it, we were faced with a difficult decision. He translated some of it to us and we saw at once that it was a masterpiece. If we published it, we should have to print at least 1,000 copies, a number which we could not possibly manage ourselves. We took the plunge and had 1,000 copies printed for us by the Pelican Press for £73. It was our first commercial venture.[1] It was an immediate success and we had to reprint another 1,000 copies before the end of the year. Kot and I translated it and I do not think that I have ever got more aesthetic pleasure from anything than from doing that translation. It is one of the most remarkable biographical pieces ever written. It makes one hear, see, feel Tolstoy and his character as if one were sitting in the same room—his greatness and his littleness, his entrancing and enfuriating complexity, his titanic and poetic personality, his superb humour. The writing is beautiful; every word and every sentence are perfect, and there is not one superfluous word or sentence in the book. I got immense pleasure from trying to translate this ravishing Russian into adequate English.

The success of Gorky's book was really the turning point for the future of the Press and for our future. Neither of us wanted to be professional, full-time publishers; what we wanted to do primarily was to write books, not print and publish them. On the other hand, our three years' experience of printing and publishing had given us great pleasure and whetted our appetite for more. In 1920 I felt in my bones that the Hogarth Press, like the universe and so many things in it, must either expand or explode or dwindle and die; it was too young and too vigorous to be able just to sit

[1] In 1919 we had had Middleton Murry's *Critic in Judgment* printed for us by a small printer, my friend McDermott, but he and I had really printed it together, and we printed only 200 copies. Virginia and I bound it.

still and survive. And we were impelled by another very powerful motive for keeping the Press in existence. Publishing our *Two Stories* and Virginia's *Kew Gardens* had shown us, and particularly Virginia, how pleasant it is for a writer to be able to publish his own books. As I have said more than once, Virginia suffered abnormally from the normal occupational disease of writers—indeed of artists—hypersensitiveness to criticism. The publisher of her first two novels was her own half-brother, Gerald Duckworth, a kindly, uncensorious man who had considerable affection for Virginia. His reader, Edward Garnett, who had a great reputation for spotting masterpieces by unknown authors, wrote an enthusiastic report on *The Voyage Out* when it was submitted to Duckworth. Yet the idea of having to send her next book to the mild Gerald and the enthusiastic Edward filled her with horror and misery. The idea, which came to us in 1920, that we might publish ourselves the book which she had just begun to write, *Jacob's Room*, filled her with delight, for she would thus avoid the misery of submitting this highly experimental novel to the criticism of Gerald Duckworth and Edward Garnett. So we decided to allow the Press to expand, if it could, into a proper publishing business, to publish a book of short stories by Virginia, *Monday or Tuesday*, in 1921, and to ask Gerald to abandon his option on *Jacob's Room* so that it could be published by the Hogarth Press.[1]

This decision to allow the Press to expand and become professional, respectable, and commercial was bound up with another major decision. Lytton Strachey was considerably intrigued by what he considered to be our eccentric publishing and printing antics. Having, in his usual way,

---

[1] Gerald agreed to this and later we purchased from him the rights in and stock of *The Voyage Out* and *Night and Day* so that all Virginia's books became Hogarth Press publications. The Press also bought from Edward Arnold the rights in my book, *The Village in the Jungle*.

poured a good deal of icy water over the head of the Press and down our backs, he began to warm up a little when we told him that we thought we should have either to kill the Press or expand it into a regular publishing business, which would mean employing someone to work with us. He suggested and was soon urging that we should take Ralph Partridge into the Press. In 1920 Ralph was a unit in a strange *ménage à trois* which inhabited a very pleasant old mill-house in Tidmarsh. It belonged to Lytton, who had just become famous by the publication of his *Eminent Victorians* in 1918. With him lived Carrington, a young woman with one of those mysterious, inordinately female characters made up of an infinite series of contradictory characteristics, one inside the other like Chinese boxes. She was the apotheosis of the lovely milkmaid who is the heroine of the song: 'Where are you going to, my pretty maid?' She had a head of the thickest yellow hair that I have ever seen, and as, according to the fashion of the time among art students at the Slade, it was cut short round the bottom of her neck, it stood out like a solid, perfectly grown and clipped, yew hedge. She had the roundest, softest, pinkest damask cheeks and large, China blue eyes through which one was disconcerted to glimpse an innocence which one could not possibly believe really to exist this side of the Garden of Eden—in 1920 in the Berkshire house of the author of *Eminent Victorians*. She was a painter, having studied at the Slade, and she habitually wore the rather sacklike dresses which in the early 1920s were worn by artistic young women and can still be seen in the works of Augustus John. For some reason unknown to me she was universally called Carrington, which was her surname; I never heard anyone call her by her Christian name and I am not quite certain whether it was or was not Doris. I liked her very much, for she was charming when we stayed at Tidmarsh or when she came to us, and always very affectionate. But she was a silent woman and rarely took part in any general conversation. It was impossible to know whether the Chinese boxes were full of intricate psychological mysteries or whether in fact they were all empty. Carrington was devoted to Lytton, running his house for him and waiting hand and foot upon him and everyone staying in the house, like a perfect housekeeper and a dedicated cook, parlourmaid, and housemaid.

Ralph Partridge was the third member of the trinity living in Lytton's house in Tidmarsh. He was a very large, very good-looking, enormously strong young man. At Oxford he had been a first-class oar and would have got his Blue if he had not suddenly taken against rowing. He fought as a commissioned officer through the 1914 war. I am not quite sure how he got to know Lytton; I think it was through Carrington with whom in 1920 he was very much in love. Ralph was an interesting character; on the surface he was typical public schoolboy, Oxford rowing Blue, tough, young blood, and on the top of this he was a great he-man, a very English Don Juan. But behind this façade of the calm unemotional public school athlete there was an extraordinary childlike emotional vulnerability. An incident made me think that Ralph's emotionalism was in part hereditary.

He once asked us whether he could bring his father to dinner with us, as he was much concerned about him and thought that conversation with us might take his mind off his worries. His worries were curious. He was, like his son, a very large man, with a surface of rugged imperturbability. He was a retired Indian civil servant. He lived in the country and Ralph from time to time went down and spent a week-end with him. Some weeks before, Ralph in his father's study had casually opened the door of a small safe and found lying in it a loaded revolver. Thinking this very strange, he asked his father why he kept a loaded revolver in an unlocked safe. At first his father tried to shrug the whole thing off, but eventually admitted that he was desperately worried. His story was this. When in India, he had bought a considerable number of shares in an Indian company. After his retirement, when he had to make out his income tax returns, for some reason or other, he had got it into his head that he was not liable to pay income tax on the dividends paid in India by an Indian company and he had not included them in his returns for many years. Then suddenly he became aware that the dividends had always been part of his income liable to income tax. He had defrauded the revenue! He wrote to the Commissioners of Inland Revenue explaining what he had done and asking them to let him know what he should now do. His letter was acknowledged and then silence from the Commissioners. He wrote again with the same result, and then a third time with the same result. He then loaded his revolver and decided that, if he did not hear from the Commissioners in ten days' time, he would commit suicide. Ralph, having induced him to hand over the revolver, wrote to the Commissioners of Inland Revenue to inform them that, if they did not reply to his father's letters, his father would shoot himself. Almost by return came a letter informing him of the total tax he must pay on the undeclared dividends. Poor Mr Partridge was saved from self-slaughter and some years later died from natural causes.

I do not think that Ralph, in a fairly long and certainly happy life, probably ever came near to suicide or even to the contemplation of it. But beneath the rather ebullient, hail-fellow-well-met, man of the world façade there was a curious stratum of emotionalism not unlike Mr Partridge's. He was easily moved to tears. He was, as I said, very much in love with Carrington. She was the classic female, if there has ever been a classic female—if the male pursued, she ran away; if the male ran away, she pursued. These tactics, applied to

Ralph, drove him into almost hysterical craziness. We decided that drastic steps must be taken. We asked Ralph whether he really, seriously wanted to marry Carrington, and he said that he did. I explained to him the phenomenon of the classic female, and told him that he must go to Carrington and put a pistol at her, not his head; he must say to her that she must marry him at once or let him go—if she said no, he would go off altogether. She gave in and married him.

Lytton, as I said, was eager that we should take Ralph into the Hogarth Press, first as an employee on trial, and with the prospect of ultimately becoming a partner. Eventually we agreed, and on August 31, 1920, the Press acquired its first paid employee. Ralph was not a full-time employee; he came and worked two or three days a week; he was paid a salary of £100 and 50 per cent. of the net profits. For the year 1920 his earnings were £56, 6s. 1d., and for 1921, £125. The first thing we had to do was to teach him to print, for his main work, at the start, was to help us with the printing. As soon as he was able to set up a page of type and machine it, we decided to develop that side of our activities: in November 1921, I bought a Minerva printing machine for £70, 10s. 0d. and 77 lb. of Caslon Old Face 12 pt. type for £18, 9s. 5d. By 1923, i.e. exactly five years after its birth, the total capital invested in the Press was £135, 2s. 3d.—all of it for printing machines, type, and materials. The Minerva machine was a formidable monster, a very heavy, treadle, platen machine, and, after treadling away at it for four hours at a stretch, from 2 to 6 in the afternoon, as I often did, I felt as if I had taken a great deal of exercise.

When the printing machine was delivered, we had it put in the corner of the dining-room, but, when McDermott saw it there, he shook his head and said it was much too danger-ous—the machine was so heavy that if we worked it there, it would probably go through the floor on to the cook's head in the kitchen. So we had to have it all dismantled again and erected in a small larder at the back of the house in the basement. The invasion of the larder was not popular with Nellie and Lottie, the cook and house-parlourmaid, but at least it was safer for them to have it there than over their heads in the dining-room.

The effect of Ralph's joining the Hogarth Press can be seen in the rapid expansion of our list to six books in 1922 and thirteen in 1923. Four in the 1922 lists were printed for us commercially. Virginia's *Jacob's Room* was our first major work, a full-length novel. 1,200 copies of it were printed for us by R. & R. Clark Ltd. of Edinburgh. This was the beginning of our long connection with one of the biggest and best of British printers and with their remarkable managing director, William Maxwell. Willie Maxwell was inside and outside a Scot of the Scots; he was a dedicated printer and a first-class business man. The moment he saw our strange, unorthodox venture into publishing, he became personally interested in it, and he took as much trouble over printing 1,000 copies for us as he did in later times over printing 20,000. When he came on his periodical business visits to the London publishers, busy though he was, he would usually find time to come out and see the Hogarth Press in Richmond. *Jacob's Room* was published in October 1922 and began at once to sell fairly briskly, and I had a second impression of 1,000 copies printed by Clark. By the end of 1923 we had sold 1,413 copies; the cost of printing and publishing up to that date had been £276, 1s. 6d. and the receipts had been £318, 6s. 0d., so that our publisher's profit was £42, 4s. 6d. We thought that we had done extremely well. It is true that Virginia Woolf, the publisher, had to some extent swindled Virginia Woolf, the author. As the whole thing was an experiment, a leap into the darkness of publishing in which we had practically no experience, and as Ralph had just come into the Press with a half share of the profits, we agreed that Virginia should not be paid a royalty, but should be paid one-third share of the profits. On the 1,413 copies sold she was paid £14, 1s. 6d.

The three other commercially printed books which we published in 1922 were Russian; they came to us through Kot, and either Virginia or I collaborated with him in the translation of them. All three were remarkable. Two of them had just been published in Russia by the Soviet Government and came to Kot through Gorky: *Stavrogin's Confession* contained unpublished chapters of Dostoevsky's novel *The Possessed* and *The Autobiography of Countess Sophie Tolstoi* had been written in 1913 by Tolstoy's wife. The other book, Bunin's *Gentleman from San Francisco*, is one of the greatest of short stories. These books, which I still think to be beautifully printed and bound, were very carefully designed by Virginia and me, and they were unlike the books published by other publishers in those days. They were bound in paper over boards and we took an immense amount of trouble to find gay, striking, and beautiful papers. The Dostoevsky and the Bunin were bound in very gay patterned paper which we got from Czechoslovakia, and the Tolstoy book in a very good mottled paper. We printed, I think, 1,000 of each of the three books and published the Bunin and Tolstoy at 4s. and the Dostoevsky at 6s. Each of them sold between 500 and 700 copies in twelve months and made us a small profit, and they went on selling until we reprinted or they went out of print.

The big expansion of the Press took place in 1923, in which year we published seven books printed by ourselves and six printed for us. *Pharos and Pharillon* by E. M. Forster, which we printed ourselves, was a terrific undertaking. It was an 80-page demy octavo volume and we printed between 800 and 900 copies. Virginia, Ralph, and I set it up, and Ralph and I machined it. It was only just possible to print four pages at a time on the Minerva printing machine, so that Ralph and I between us had to treadle 22 runs of over 800 pages a run. The first edition sold out in less than a year; the receipts, at a published price of 5s., were £135, 10s. 11d., and our expenditure £90, 19s. 0d., so that the book showed a profit for the Press of £44, 11s. 11d. We at once had a second edition reprinted for us, crown octavo, and published it in paper covers at 3s.

The three of us must have done a tremendous amount of printing in the years 1921 and 1922 in order to produce the crop of our hand-printed books published in 1923. For they included *The Waste Land*, 37 pages, and Robert Graves's *The Feather Bed*, 28 pages crown quarto; a large book of Roger Fry's woodcuts, which was not easy printing for amateur novices; two crown quarto books of poetry by Herbert Read and Clive Bell, the latter being illustrated and decorated by Duncan Grant; and a small book of poems by Ena Limebeer. We bound the woodcuts and Ena Limebeer's poems ourselves, but the other books were too large for us to tackle ourselves and we had them bound for us by a commercial bookbinder.

The Hogarth Press, in these early years, met with a rather chilly welcome, or rather cold shoulder, from the booksellers. If you compare the thirteen books which we published in that year with any thirteen similar books from other publishers, you will find that all of ours have something more or less unorthodox in their appearance. They are either not the orthodox size or not the orthodox shape, or their binding is not orthodox; and even worse, what was inside the book, what the author said, was in many cases unfamiliar and therefore ridiculous and reprehensible, for it must be remembered that, if you published 42 years ago poetry by T. S. Eliot, Robert Graves, and Herbert Read and a novel by Virginia Woolf, you were publishing four books which the vast majority of people, including booksellers and the literary 'establishment', condemned as unintelligible and absurd. Conservatism is the occupational disease in all trades and professions, and booksellers suffer from it like everyone else. In 1923 we had no travellers and in a very desultory way we took our books round to the more important booksellers ourselves in order to get subscription orders before publication. It was a depressing business, though no doubt salutary and educative for an embryonic publisher. There were a few booksellers, like the great Mr Wilson of Bumpus, James Bain of King William Street, Lamley of South Kensington, Goulden and Curry in Tunbridge Wells, the Reigate bookshop, who were immediately interested in what we were trying to do and did everything to help and encourage us. But they were the exception. The reception of *Jacob's Room* was characteristic. It was the first book for which we had a jacket designed by Vanessa. It is, I think, a very good jacket and today no bookseller would feel his hackles or his temperature rise at sight of it. But it did not represent a desirable female or even Jacob or his room, and it was what in 1923 many people would have called reproachfully post-impressionist. It was almost universally condemned by the booksellers, and several of the buyers laughed at it.

Most human beings will never move unless a carrot is dangled in front of their noses, and, like the donkey, they must have precisely that kind of carrot which they and their fathers and fathers' fathers back to the primal ass have always recognized as the only true, good, and respectable carrot. Our books 42 years ago were not recognized by the trade as the right kind of carrot, either internally or externally. But looking at them today any bookseller would admit that they are extremely well-produced books and that their jackets are admirable. Within ten or twelve years the binding of books in gay, pretty, or beautiful papers over boards was widely adopted for all kinds of books, particularly poetry. Time or the rise and fall of reputations have justified our judgment of the inside of these books. Only three out of the thirteen, Ena Limebeer's *Poems*, Mrs Lowther, and Reynolds's letters, would not be recognized today as important books by important writers, and even for these there is still something to be said. As for the other ten, there are few publishers who in 1965 would not be very glad to have them on their lists.

In 1922 a storm blew up in the Hogarth Press, the kind of crisis which, during the next 20 years, was to recur with depressing regularity. After two years' experience it was clear to us that our arrangement with Ralph was not turning out successfully. He worked only two or three days a week, and that rather erratically. It was an impossible arrangement if we were to publish, as we did in 1923, twelve books in the year, and ourselves print half of them. We wanted Ralph to become full time, a professional publisher. This he would not do, though he was enthusiastic about the Press and, with tears in his eyes and voice, maintained that nothing would induce him to give it up.

The truth was—as I suspected at the time and now see clearly looking back to 1923—that we were trying to do what is practically impossible, enjoy the best of two contradictory worlds. The success of the Press was forcing it to become a commercial publishing business. My experience in Ceylon had taught me (I think immodestly) to be a first-class business man, but I was not prepared to become a professional publisher. The Press was therefore a mongrel in the business world. We ran it in our spare time on lines invented by myself without staff and without premises; we printed in the larder, bound books in the dining-room, interviewed printers, binders, and authors in a sitting-room.

I kept the accounts, records of sales, etc., myself in my own way, which was from the chartered accountant's view unorthodox, but when it was challenged by the Inland Revenue and I took my books to the Inspector of Taxes, he agreed that they showed accurately the profit or loss on each book published, the revenue and expenditure of the business, and the annual profit or loss, and for many years the Commissioners accepted my accounts for the purpose of assessing income tax.

The organization and machinery of the Press were amateurish; it was, so far as Virginia and I were concerned, a hobby which we carried on in afternoons, when we were not writing books and articles or editing papers. We did not expect to make money out of it. But at the same time we were already committed to publish full-length, important books, not only for Virginia herself, but also for writers whom we considered important. We felt to these writers and their books the responsibility of the commercial publisher to the author, we had to publish their books professionally and competently. Our idea in 1922 was to get someone like Ralph who would work full time in the Press under me and earn his living from it, while Virginia and I would continue to work at it in our spare time as a by-product of our life and energies.

The last sentence shows what a very curious type of business we were trying to create and that the position of the young man who was to work under me would not be easy— nor would it be easy to find the right young man. I have never been an easy person to work with. I emphasize the words 'work with'. My experience in the Ceylon Civil Service proved that I get on much better with subordinates than with equals or superiors in business. In practical affairs I am in many ways a perfectionist—a character for which in the abstract, or when I see it in other people, I have no great admiration. I have a kind of itch or passion for finding the 'right' way of doing things, and by 'right' I mean the quickest and most accurate and simplest way. In 1923 I was still young enough to be hot tempered and allergic to fools.

By the middle of 1922 it was clear to us that Ralph, from our point of view, would not do. As it was, I was becoming a full-timer and he a part-timer, whereas we wanted the exact opposite. We found ourselves sitting uncomfortably on the horns of a dilemma—and the same situation would build itself up again and again from time to time over the following years. Shall we give up the Press altogether or shall we make one more attempt to find a manager or a partner who will help us to run this commercial hippogriff on the lines on which we want it to develop? We were then (and more than once again in the future) very much inclined to give the whole thing up and leave ourselves free of responsibility to pursue our other activities. On the other hand, we were urged from the outside to develop the Press and naturally were rather flattered by this. James Whittall, a cultured American, put out feelers and we considered him as a possible partner. But more surprising was a direct offer from the great publishing house of Heinemann to take us into a kind of partnership or 'association'. We had several talks with Whittall and, invited by the managing director of Heinemann, I went and had an interview with him on November 27, 1922. He offered to take over the whole business management of the Hogarth Press, distribution, accounting, advertising, and, if we wanted it, printing and binding. We should be left complete autonomy to publish or not to publish any book we liked.

We turned down Whittall and we turned down Heinemann. We liked Whittall very much personally, but we came to the conclusion that he was too cultured for us and for the Press. We did not want the Press to become one of those (admirable in their way) 'private' or semi-private

Presses the object of which is finely produced books, books which are meant not to be read, but to be looked at. We were interested primarily in the immaterial inside of a book, what the author had to say and how he said it; we had drifted into the business with the idea of publishing things which the commercial publisher could not or would not publish. We wanted our books to 'look nice' and we had our own views of what nice looks in a book would be, but neither of us was interested in fine printing and fine binding. We also disliked the refinement and preciosity which are too often a kind of fungoid growth which culture breeds upon art and literature; they are not unknown in Britain and are often to be found in cultivated Americans. It was because Whittall seemed to us too cultured and might want to turn the Hogarth Press into a kind of Kelmscott Press or Nonesuch Press that we turned him down. I am, of course, aware that many people would have thought—and some would still think—it ludicrous for us, and particularly Virginia, to talk of anyone being too cultured. The myth of Virginia as queen of Bloomsbury and culture, living in an ivory drawing-room or literary and aesthetic hothouse, still persists to some extent. I think that there is no truth in this myth. Her most obvious fault, as a person and as a writer, was a kind of intellectual and social snobbery—and she admitted it herself. There is also sometimes a streak of incongruous archness in her humour which is almost ladylike and very disconcerting. But her novels, and still more her literary criticism, show that she had not a trace of the aesthete or hypercultured. One has only to compare her attitude towards life and letters, towards art and people, with that of writers like George Meredith or Henry James or Max Beerbohm, to see that, although she was a cultured woman, the roots of her personality and her art were not in culture and that she had a streak of the common-sense, down to earth, granitic quality of mind and soul characteristic in many generations of her father's family.

We turned down Heinemann for the simple reason: *timeo Danaos et dona ferentes*, I fear the Greeks, especially when they offer me gifts. We felt that we were really much too small a fly to enter safely into such a very large web. So there we were with two pretty heavy albatrosses hanging round our necks—the Hogarth Press and Ralph Partridge. Lytton and Ralph put forward various proposals which we could not agree to, and sometime in 1922 we more or less agreed to part. Then a curious thing happened. In November 1922 we were in the midst of our negotiations and conversations and hesitations, and on the 17th of that month we met Whittall in the 1917 Club to discuss Heinemann's offer with him. While waiting for him, a young woman came in and began talking to a man who was sitting near us. She is described in Virginia's diary as 'one of those usual shabby, loose, cropheaded, smallfaced bright eyed young women'. It was impossible not to overhear the conversation. She told the man (who, I think, was Cyril Scott, the composer) that she was sick of teaching and had decided to become a printer. 'They tell me,' she said, 'that there's never been a woman printer, but I mean to be one. No, I don't know anything about it, but I mean to be one.' Virginia and I 'looked at each other with a wild surmise', and when the young woman went out, Virginia followed her and brought her back to our table. There we told her what we were doing with the Press and arranged that she should come and see it for herself and discuss possibilities.

Some days later on a Sunday afternoon the young woman, whose name was Marjorie Thomson, accompanied by a friend, came and had tea with us at Richmond in order to see what we were doing and discuss her possible employment. This was the first time I met Cyril Joad, for the friend

81

was Joad, who was to become famous later as Professor C. E. M. Joad, the highbrow radio star. Cyril was a curious character; high minded, loose living and loose thinking, he inhabited a kind of Platonic or Aristotelian underworld. He was one of those people whom I dislike when I do not see them and *rather* like when I do see them. He was in fact a selfish, quick-witted, amusing, intellectual scallywag. He told us that he was going to marry Marjorie in a few months' time, and in a few months' time Marjorie told us that she had married him. They lived together for some time, not I think very happily, in the Vale of Health, but the marriage was one of the many figments of Professor Joad's fertile imagination.

Marjorie had a nice face and a nice character. She belonged to what Virginia called the underworld, and it was, I suppose, looked at from certain altitudes, both socially and intellectually an underworld, a twentieth-century mixture of Bohemia and Grub Street; a certain number of its inhabitants could always be met in the 1917 Club in Gerrard Street. Marjorie had a bright, if somewhat shallow, mind, and as soon as she saw what we were doing in the Press, was anxious to join us. With Professor Joad's benediction, it was agreed that she should come to us on January 1, 1923, on a salary of £100 and a half share of the profits, Ralph leaving the Press finally in March. This, then, was the position of the Press when in March 1924 we moved from Hogarth House, Richmond, to Tavistock Square in Bloomsbury, and a new phase with rapid development began for our publishing. I will deal with this in the next chapter.

Our energy in the last four years of our life in Richmond was considerable. With no staff, to publish thirteen books (seven printed by ourselves) in a year would, I think, have been considered by many people a full-time job. I kept all the accounts myself and did a good deal of the invoicing. This in itself was no light labour. For instance, *Jacob's Room* was published on October 27, 1922, and by the end of 1923 it had sold 1,182 copies on over 200 orders. Each of these orders was entered by me in a serial number book and in a ledger; the invoices were made out either by me or by Ralph; and the books were packed and despatched by Virginia, Ralph, and me. This was our spare-time occupation almost always confined to the afternoon. During the same twelve months of 1923 Virginia was writing *Mrs Dalloway*, preparing *The Common Reader*, and earning £158 by reviewing. My own major occupation or occupations had become more complicated and variegated.

I wrote *Empire and Commerce in Africa* in 1918 for the Fabian Society. It is a formidable book of 374 pages and I did a great deal of intensive reading for it. It is, I think, one of the earliest studies of the operations of imperialism in Africa. In 1920 Philip Snowden asked me to contribute a volume to the Independent Labour Party's 'Social Studies Series'. In the early 1920s there was a triumvirate of Labour Party leaders, Ramsay MacDonald, Arthur Henderson, and Philip Snowden. Snowden, whom I never knew as well as I knew the other two, was a curious man. He was lame and gave one the impression of being embittered by pain, though in ordinary conversation it was a rather gentle embitterment. I was an active member of the I.L.P., which in those days was the left wing of the Labour Party, and we were inclined to believe that Snowden was the most advanced or progressive of the triumvirate. In this we were very much mistaken. In *Beginning Again* (pp. 217-226) I have dealt with Ramsay; he was an opportunist who genuinely confused the highest political principles with the personal interests of James Ramsay MacDonald; he was neither on the Left nor on the Right, he was always bang on the Centre, and the Centre was James Ramsay MacDonald. In those early days I underestimated Henderson, thinking him to be what he looked like and what his political petname of Uncle Arthur seemed to indicate—a rather stuffy, slow-going and slow-thinking professional politician. He was something more and a good deal better than that. He was a man of some political principle and political understanding, a rare phenomenon among Cabinet Ministers anywhere. He was still capable of genuine and generous feelings for what he thought politically or socially right, and the desiccation of years in the trade union movement and the Labour Party had not dimmed or dulled these feelings. On the surface these were concealed behind a slow, watchful, slightly suspicious stare from hooded eyes, not uncharacteristic in those days of the British working man. Henderson's calibre was shown by his work as President of the League of Nations Disarmament Conference in 1932 and his behaviour to the great rat race of 1931 when Ramsay and Snowden temporarily destroyed the Labour Party. He had been himself chairman of the Labour Party before 1919, and the history of the party, Britain, and even Europe might well have been different and less catastrophic if he and not Ramsay had been leader and Prime Minister in the 1920s.

Snowden, the third member of the triumvirate, was quite unlike Henderson—and indeed quite unlike Ramsay. He came, I think, from the lower middle class and had been a subordinate civil servant and journalist. He was really an old-fashioned Liberal, which meant that by the time that I knew him he was in most things, from my point of view, about as progressive as a member of the Junior Carlton Club whose political faith was limited to support of the Crown, the Church, and free trade. He was one of those very honest, unimaginative, conservative—fundamentally reactionary—politicians who drift into a left wing, progressive, or even revolutionary party, and do a good deal of harm politically. That he and Ramsay were leaders of the Labour Party in

the crucial years from 1919 to 1931 was a disaster not only
for the party but also for Britain, for their leadership inevitably landed them and all of us in the barren wilderness of the
1930s and the howling wilderness of the war. The most
dangerous thing for a boat in a stormy sea is to find herself
with no rudder; thanks to Ramsay and Snowden, that was
the condition of the Labour Party in the years which followed
the 1914 war. Though the upper ranks of the party have
always been full of intellectuals, Labour has always shared the
general British suspicion and misprision of the intellect and
of those who use it in everyday life. As an unredeemed and
unrepentant intellectual I was only too well aware of the
widespread feeling that intelligence, unless camouflaged by
silliness or stupidity, is dangerous and discreditable, and I
never felt comfortable with Snowden who was a very British
Briton and did not, it seemed to me, like me or my intellect.
I was surprised when he asked me to write a volume for the
I.L.P. series. I wrote a short book with the title *Socialism
and Co-operation*, for which I was paid £25.

I have always been a heretical socialist, and, since very few
heresies ever become orthodoxies, this book was even more
futile than most of my books. Yet I still think that what the
book said is both true and important, though neither the
true-blue capitalist nor the true-red socialist nor even the
pinkish trade unionist will have anything to do with it. The
gist of my argument was that, in the modern world, socialism, i.e. the ownership or control of the means of production
by the community, is not an end in itself, but an essential
means to a prosperous and civilized society, and that the
ownership and control should be based not on the state or
the organized producers, but on the organized consumers.
The curse of the capitalist system is that it produces states
of mind in individuals and classes which contaminate society
by inducing a profound, instinctive conviction that the object

and justification of everyone's work, trade, profession, in fact of nine-tenths of a person's conscious existence from youth to senility, are and should be money, i.e. the personal interests of the individual. The machinery of communal economics and production is organized not to produce what the community wants or needs to consume but in order to provide either profit or a salary or wage or just 'work' for individuals. Marxist socialists and all those variegated 'packs and sets' of communists and socialists who, since Marx, 'ebb and flow by the moon', because they start from the exploitation of the worker, have unconsciously accepted the psychology of capitalism; both in theory and, where they have the opportunity, in practice, they organize society in the interest of the producers, not the consumers, the economic system being geared to and judged by its ability to provide work, wages, and salaries, competition for the profits between different classes of worker being substituted for competition between capitalists. This process has, of course, been enormously encouraged because in Europe trade unionists and trade unionism have been a dominating influence in all socialist parties, and the trade unionist, as a trade unionist, is concerned not with what he produces and its consumption but with 'work', which the production provides for him, and the amount of money which he can make out of it.

The importance of the British co-operative movement and system is that they have proved that efficient control of large-scale production and distribution by consumers is possible. I argued that it was possible and desirable to develop and extend this system into consumers' socialism, i.e. the control of the industrial system by the community organized as consumers, and that this would not only revolutionize the whole economic system but also the social psychology of capitalism and of socialism based upon production and the interests of the producers.

I think that the nature of social or economic organization has an immense effect upon social psychology for good or ill. I will give a trivial example which, nevertheless, always seems to me to throw light upon our civilization and upon the barbarous psychology which it has bred in us. If you stand at any bus stop in London and observe what happens to the queue of consumers, i.e. of the people who wish to use the bus to get from one place to another in London, you will note a strange thing about this simple industrial or economic operation. The buses drive up very often in conglomerations of three, four, or even five, all of the same number, i.e. all going to the same place—the first bus being full, the second half-full, and the two or three others empty. They dash off almost immediately and the slightest hesitation on the part of anyone in the queue will leave him behind to wait a long wait for another conglomeration of buses. The explanation of this is that the system of transport by bus in London is not organized primarily with the object of transporting the public as rapidly and comfortably and efficiently as possible from one place to another, i.e. in the interests of consumers, but for three entirely different objects: first, for what is known as a schedule, i.e. the timetable which says that the bus must leave a particular garage at a particular time and arrive at another garage at a particular time; secondly, to provide work for the drivers and conductors, i.e. to provide the highest possible wage under the best possible working conditions for the producer; thirdly, to obtain from the public the maximum amount in fares so that, while paying the highest possible wages, the whole business can be run at a profit, or at least not a loss. The interesting fact is that this barbarous system is accepted as rational, inevitable, and civilized, not only by the management, the major and minor bureaucrats who direct London Transport, not only by the workers and employees, but even by the under-dogs in the queue, the cringing consumers. And members of the Labour Party and the trade unions will welcome this system as socialism, provided that it is controlled at the top by a Transport Board instead of a Board of Directors.

The psychology of trade unionism and the psychology of capitalism, which spring from the same social postulates, and are complementary like the positive and negative in electricity, are so firmly established in modern society that to most people the idea that the object of industry should be consumption, not production or profit, seems Utopian and even immoral. I never imagined, therefore, that my argument in favour of socialism controlled by consumers would cut any ice in the Labour movement. But even in politics, where reason is so suspect and so unwelcome, I have an absurd, pig-headed feeling that one ought to use one's reason. However, I propose to leave the account of my political activities to the next chapter and to return now to my other occupations in the years 1919 to 1924.

Between 1921, when *Socialism and Co-operation* was published, and 1931, when *After the Deluge*, Vol. I, was published, I did not produce any book, partly because I was so much occupied by journalism, publishing, and politics, and partly because I was ruminating and slowly writing *After the Deluge*. But in 1921 I had one incident connected with my writing which amused Virginia and me. In that year we published in the Hogarth Press a book of three short stories by me, *Stories from the East*, and one of them, 'Pearls and Swine', said Hamilton Fyfe in a review in the *Daily Mail*, 'will rank with the great stories of the world'. Mr Henry Holt, a literary agent, on the strength of this review wrote and asked me whether I would let him have the story for America. I did not do so, and months later in 1922 he wrote to me again, asking whether I would let him deal with the story, which 'ought to have netted you anything up to a couple of hundred pounds for serial rights. Possibly more'.

This time I sent him the book—I don't think that up to this time he had read the story—and told him that he could try his hand. When he read the story, it was obviously a bit of a shock to him, being a good deal too plain spoken for the two hundred pound bracket in the United States of America. He wanted me to tone it down a bit—he called this euphemistically 'a few artistic alterations'—and send it to the great American literary agent, Ann Watkins, who had already expressed an interest in it. I replied that I 'cannot bear to contemplate rewriting anything which I wrote a long time ago', but that he could deal with the story and Ann Watkins himself. This he did. Ann Watkins also thought the story a masterpiece, but was also obviously horrified by it and the idea of offering it to the American market. 'The realism, the vivid picture quality of "Pearls and Swine",' she wrote, 'is so great as to be terrific. It is as powerful a story as I have read in a long time. . . . But there are only about two magazines in America that I think would touch it. You see, we here in the States are still provincial enough to want the sugar-coated pill; we don't like facts, we don't like to have to face them. It seems to be a characteristic of the American people. And where we won't face them in our politics, in our domestic problems, in our personal lives—why in the devil should we be forced to face them in fiction? I think, fundamentally, our demand of the author is that he entertain us with his wares. We veer from the shocking, the revolting—the truth. But holy, suffering cats! how Woolf can write! I should like nothing better than to represent him in the American market. But I should like to represent him only if in so doing I can be of profit to both him and myself.'

Mr Holt was much impressed and wrote to me: 'I wish I could make you realize the tremendous commercial significance of this'. He wanted me to settle down to writing something suitable for the great American market—'I may never again.' he said, 'have patience to bully a man into making several thousand a year, so, for the last time, *do think it over*'. I do not think that I answered this letter, and then one afternoon a car stopped at the gate of Monks House, and out of it stepped Mr and Mrs Holt. He said that he wished to talk to me alone while Mrs Holt would talk to Virginia. It took me the better part of an hour to get rid of him. He said that I must devote myself to writing stories with him as my literary agent, and that, if I did, he would guarantee that I should make £3,000 a year. I said that I didn't want to write stories and in any case I could never think of plots for them. Eventually we went in and joined the ladies over a cup of tea. Virginia recorded the following conversation: ' "He sells everything—he'll be selling me next," she says, very arch. Mr Holt half winked and cocked his head. "Little woman, little woman," said Mr Holt. "He's the straightest boy that ever lived," said Mrs Holt, not without emotion.' At last they drove away, and some days later I got a letter from Mr Holt in which he gave me the outline of a plot for a short story which I should write, and he assured me that, if I would do so and send it to him, he would get me a large sum of money for it. I did not write Mr Holt's story and I do not think that I ever heard from him again.

I have told in *Beginning Again* how it came about that I started and edited the *International Review* for the Rowntrees. It did not last very long, for Arnold Rowntree had, I think, underestimated the costs and loss involved in financing a monthly review dealing with international affairs. When it came to an end, Rowntree asked me to write regularly 16 pages on international affairs for the *Contemporary Review*, which was also a Rowntree paper. I did this in 1920 and 1921 for the noble fee of £250 per annum. From the great newspaper proprietors, like Lord Northcliffe and Lord Beaverbrook, down to the humble sub-editor or reporter, and the still more humble writer for the highbrow weeklies, monthlies, and quarterlies, it is almost impossible for the journalist and the owners of journalists not to believe that what they write or what they hire other people to write has great influence and importance. In general, the bigger the journalistic bug, the bigger his delusion about his influence and importance. I have no doubt that no one who writes for papers ever completely sheds this delusion and that I myself still nourish in my unconscious a secret and sheepish hope, if not belief, that a few people will be influenced by what I write. But I think that my early journalism—writing for the *New Statesman* and for the *Nation* and editing the *International Review*—rapidly disillusioned me. In writing, it seems to me, one just has to cast one's bread upon the waters, resigning oneself to the fact that nothing will ever return to you except so many pounds per thousand words,[1] and the soggy bread will sink without a trace. Certainly one would have had to be very artless or very sanguine to think that many people read or anyone minded 16 pages on international affairs in the *Contemporary Review*. I found it a depressing job, and was not sorry to hand it over to George Glasgow in 1921.

In 1920 and 1921 I still did quite a lot of reviewing and article writing for the *New Statesman* and the *Nation*, earning from the *New Statesman* £65, 7s. 0d. in 1920 and £44, 4s. 0d. in 1921, and from the *Nation* £81, 8s. 0d. in 1920 and £40, 14s. 0d. in 1921. The large sum of £81 which I earned from the *Nation* in 1920 was due to the fact that, as I recorded in *Beginning Again* (p. 185), for three months during that year I temporarily took H. N. Brailsford's place as leader writer on the paper. The *Nation* was owned by the Rowntrees, the Quaker chocolate and cocoa kings of York,

[1] 'No man but a blockhead ever wrote except for money,' said Johnson.

and was edited by H. W. Massingham, one of the most famous editors of his time. In 1922 Brailsford became editor of the *New Leader* and Massingham asked me to take his place on the staff of the *Nation*. I accepted and landed myself in a tangle of events and a journalistic career which lasted for eight years. Initially my duties when I took over Brailsford's job were to go to the *Nation* office in Adelphi Terrace every Monday morning and arrange with Massingham what I should write for next Saturday's paper. Practically always it consisted of the first leader on some political subject, three or four notes on political subjects, and occasionally a review. Massingham was a strange, rather disquieting person. He was a small, neatly dressed, quiet-spoken man whose face had the look of one of those small, brindled, reserved mongrels who eye one with motionless suspicion—the expression of eye and mouth always fills me with apprehension. In 1922 he was 62 years old and had edited the *Nation* for 15 years. All his life he had been a pillar of both liberalism and Liberalism and his whole journalistic career had been on Liberal papers; before editing the *Nation* he had been editor of the *Daily Chronicle*. The *Nation* was supposed to be, and had been, a Liberal weekly supporting the Liberal Party, and Arnold Rowntree, the head of the Rowntree clan, was a Liberal M.P. These facts are important for an understanding of what happened in the next twelve months and of the entanglements in which to my surprise I found myself. For Massingham had in the years just before 1922 drifted further and further from the Liberal Party and had become gradually a supporter of the Labour Party whereas the Rowntrees remained Liberals.

I did not realize any of this when I first stepped into Brailsford's shoes. It was a peaceful office with the gentle, deaf H. M. Tomlinson as literary editor and Miss Crosse one of those highly geared, super-efficient secretaries who are themselves capable of editing and often, *de facto* but not *de jure*, do edit the paper. Then there was a band of very distinguished, veteran Liberals, old friends of Massingham's, who formed the staff of the paper, J. A. Hobson, J. L. Hammond, and H. W. Nevinson. In their company I felt very much the new boy and they were so high-minded—the particular brand of high-mindedness seemed to be peculiar in those days to Liberals who lived in Hampstead and Golders Green—that I always felt myself to be a bit of a fraud in their company. Hammond and Hobson were two charming men; I liked them very much and I became a friend of both; to many people Nevinson was a great charmer, but he was altogether too noble for my tastes. I don't like knights *sans peur et sans reproche*—they, like mongrels, make me uneasy. These three with Massingham and Tomlinson used to lunch together on Mondays and I often used to join them.

Before the lunch I had had my talk with Massingham and decided with him what I was to write. I never felt that I really understood Massingham. He was always extremely nice to me and I got on with him very well, both in work and over the lunch table, but I do not think that I ever got more than a fraction of an inch below the surface. The routine of Monday morning was that he invariably asked me to suggest the subject of my article and notes and left it to me to tell him what I proposed to say. I do not remember him ever not accepting my subject or line of policy, and his comments and suggestions were always very few. But though he said very little when we were talking as professionals on the week's job, there is no doubt that he was a first-class editor in that somehow or other he impressed his personality on those who wrote for him and what they wrote. The consequence was that the paper too had a personality, a flavour, a smell of its own, and this got into what one wrote when one was writing for it. I was never conscious of writing differently in the *Nation* and the *New Statesman* and in my own books—I never felt Massingham looking over my shoulder or breathing down my journalistic back—but if I reread what I had written for him, I was startled to get a faint whiff of Massingham and Massingham's *Nation*.

His editorial and political personality and odour or aura were complex and fascinating. First, he was extremely high-minded; the political aura of the *Nation* in 1922 was still that of Gladstonian liberalism impregnated with sophisticated or civilized non-conformity. Secondly, he was a gentle man, on the side of culture and quiet, of sentiment, if not sentimentality; the *Nation* again reflected this, being pacific, humane, with occasionally a tear—some people said a crocodile's tear—in its eyes and voice. Thirdly, he was a bitter and violent man, with a peculiar bitterness of which I will say more in a moment; the *Nation* had an undercurrent of aggressive acerbity and sudden bursts of intemperance.

I found the study of his public and private character and its extraordinary contradictions absorbing. After we had decided what I should write, we nearly always had for ten minutes or more a general conversation on the political situation. It was a time of continual crisis: Lloyd George's fatal adventure in the Middle East, the Conservative revolt against him, and the break up of the coalition government. I suppose that Massingham must have been in pre-war years a fervid supporter of Lloyd George's, but when I knew him, he hated him with an almost crazy violence and bitterness. He was, as I said, a gentle, quiet-spoken man, but nine Mondays out of ten he would begin a tirade against L. G. He never raised his voice, but out of his mouth poured a kind of commination recital against L. G. and, frequently linked to him, J. P. Scott, editor of the *Manchester Guardian*. I do not know why he had conceived such violent hatred of the immaculate Scott, the journalistic pillar of liberalism, but he astonished me by his venomous and grotesque accusations. I could scarcely believe that I was not dreaming a mad dream when I heard him more than once accuse these two spotless Liberals of a homosexual passion for each other.

Not much has been written about the psychology or psychopathology of political beliefs and emotions. I can remember only two books of any importance. The distinguished psycho-analyst Edward Glover in *War, Sadism and Pacifism* maintained that one had to be peculiarly sadistic and bloody-minded to be a pacifist or even a supporter of the League of Nations, and *Personal Aggressiveness and War* by Evan Durbin and John Bowlby was another original book on more or less the same subject. When I first read Dr Glover's book, I thought that his own unconsciousness was not altogether unconcerned with his finding such very discreditable motives for pacifism in the unconscious of the pacifist, but I have no doubt that there was a good deal of truth in his main thesis. I am sure that if one could look deep into the minds of those who are on the Left in politics (including myself), Liberals, revolutionaries, socialists, communists, pacifists, and humanitarians, one would find that their political beliefs and desires were connected with some very strange goings on down among their ids in their unconscious.

At any rate, watching and listening to Massingham on a Monday morning, I often felt that something like this could only explain the conflict in his character, his gentle high-mindedness and absurd verbal violence. Down below he was, I think, a man of strong feelings which might range from the milk of human kindness to hatred and bitterness. It is also possible that he had the diffused dissatisfaction and grudge against the universe that small men often have, though I may be wrong about this and indeed about his stature. At any rate, as Freud insisted in *Civilization and its*

Discontents, to be even a moderately civilized man is not only difficult but also extremely painful. If you have to be as high-minded all the time as a Liberal of Golders Green or Welwyn Garden City, editing or writing for the *Nation*, in 1922, you had to be suppressing all the time some very violent and curious instincts which might, I think, have surprised and shocked even the editor of the *Nation* had he found them in his unconscious mind. Nineteen hundred years before Freud wrote *Civilization and its Discontents* Horace said *naturam expellas furca, tamen usque recurret* (you may drive nature out with a fork, but she will always return) —and we now know that she returns in strange and very different forms. I have no doubt that the strain of being so civilized caused the explosions of Massingham's verbal violence.

His queer, secretive, complex character had, I am sure, much to do with causing the curious situation in which I found myself involved about a year after I joined his staff. The Rowntrees, proprietors of the paper, were, as I said, Liberals and the *Nation* had been, and was still supposed to be, a Liberal paper. But in fact, when I joined it, it was to all intents and purposes a Labour paper. Massingham had become bitterly hostile to the Liberals and Hammond, Hobson, Nevinson, like Brailsford, Noel and Charles Buxton, who before the war had been active and distinguished Liberal intellectuals, had all drifted into the Labour Party. Massingham never told us exactly what happened behind the scenes between him and the Rowntrees. I am inclined, knowing him and them, to believe that Arnold Rowntree treated him very well, allowing him a great deal of latitude, but warning him that the Rowntrees could not agree to the paper becoming a mere Labour Party organ. All through 1923 the divergence between their political views increased and suddenly there was an explosion or showdown.

I had been told nothing of what was going on, and I was astonished when one morning Massingham told me that the Rowntrees had decided that they could not go on financing the paper as it was, and that they had decided to sell it, but would give him first option to buy it. Massingham said that he knew someone who would put up the money, and he asked me whether I would agree to continue on the staff under a new proprietorship; Hammond, Hobson, Nevinson, and Tomlinson had all agreed to stand by him. I said I would.

There followed weeks of unpleasant doubt and mystery. He was understood to be hard at work raising the money, but he told us nothing. At one moment he went off to the South of France to deal with the man who was to supply most of the money, and we were left to carry on with the editing of the paper. We met for dejected lunches and speculated gloomily on what could be happening. Then Massingham returned, but told me nothing definite. I was astonished when Maynard Keynes in March 1923 told me that he and some others had acquired an interest in the *Nation*, the Rowntrees also still retaining an interest. Hubert Henderson was to be editor and Maynard asked me to become the literary editor. I went to Massingham and explained to him what had happened. I said that if he was going to acquire the *Nation* or was going to be editor of a new weekly, I would continue with him, but, if there was to be no weekly edited by him, I would accept Maynard's offer, provided that he had no objection to my doing so. He said that he had failed to raise the money to purchase the *Nation* and there was no immediate prospect of his editing a new weekly so that I had no longer an obligation to him and I must be free to accept Maynard's offer. This I did, but I insisted upon the following two conditions, which were accepted by Maynard and by Hubert: (1) I would do the work in my own time, though I would normally come to the office on Mondays,

Tuesdays, and the morning of Wednesdays; (2) I should be autonomous in my part of the paper, though the editor would have the right to object to and require the removal of anything which I had passed, but, if he did, I would then have the right to insist that Maynard should arbitrate between us. Maynard only once had to arbitrate between us during the seven years in which I was literary editor. I had given a book to review to David Garnett, then at the beginning of a distinguished literary career, but not yet, I must admit, a very skilful reviewer. When Hubert read his review in proof, he said that it was not good enough and required me not to print it. I insisted that Maynard should arbitrate, as the review, though not very good, did not merit rejection. Maynard agreed with me and the review was published.

The events, such as they were, of my seven years' work on the new *Nation*—which became, of course, a Liberal paper, for Maynard and Hubert were both Liberals—belong to the next chapter and I will leave to that chapter too my political activities on the Labour Party committees and the Fabian Society. In the four years 1920 to 1923, as Virginia's health grew more stable, our social life increased and became more and more of a problem. Our taste in human beings was pretty much the same, but we did not always agree about the best way of seeing them. Virginia loved 'Society', its functions and parties, the bigger the better; but she also liked—at any rate in prospect—any party. Her attitude to this, as to most things, was by no means simple. The idea of a party always excited her, and in practice she was very sensitive to the actual mental and physical excitement of the party itself, the rise of temperature of mind and body, the ferment and fountain of noise. Sometimes she enjoyed it as much in the event as in anticipation, and sometimes, of course, owing to her peculiar vulnerability to the mildest slings and arrows of (not very) outrageous fortune, she would leave a boring party in despair as if it were the last scene of Wagner's Götterdämmerung with Hogarth House and the universe falling in flames and ruin about her ears. Of one of these catastrophic depressions in August 1922 she wrote in her diary: 'No one ever suffered more acutely from atmosphere as I do; and my leaves drooped one by one; though heaven knows my root is firm enough. As L. very truly says, there is too much ego in my cosmos'.

She not only enjoyed society, the kaleidoscope of human beings, conversation, the excitement of parties, she was through and through a professional novelist, and all this was the raw material of her trade. This dual sensitivity to the most trivial meetings with her fellow human beings meant that society and parties were a great strain on her mental health and she herself was well aware of this. The following is another extract from her diary in the summer of 1922: 'Clive came to tea yesterday and offered me only the faded and fly blown remnants of his mind. He had been up late. So had I—at the pictures. For my own part, all my strings are jangled by a night out. Dissipation would rot my writing (such as it is, I put in, modestly). Words next day dance patterns in my mind. It takes me a week to recover from Lady Colefax—who by the way invites me for Friday'.

Virginia always thought she was going to enjoy a party enormously before she went to it and quite often she did. I did not share her optimism, nor, therefore, ever quite so keenly her disappointments, and, though I sometimes enjoyed parties, I never felt the exhilaration which they sometimes gave to her. When we were still living in Richmond, she wrote in her diary that she and I were becoming celebrities and that I denied this, but then I had not, as she had, gone to Logan Pearsall Smith's tea-party in Chelsea or to the week-end with Ottoline Morrell at Garsington. I did occasionally go to Logan's tea-parties where one drank

Earl Grey's china tea amid china, furniture, pictures, books, and human beings, not easily distinguishable from one another or from the tea with its delicate taste and aroma, for they were all made, fabricated, collected in accordance with society's standards of sophisticated culture and good taste. Earl Grey has never been my cup of tea, nor was Logan.

I occasionally went to Garsington, but not as often as Virginia. Garsington, its week-ends and Ottoline and Philip, have been described, with or without venom, in many memoirs and novels, and I have myself had something of a say in *Beginning Again* (pp. 198-203), and I do not propose to say much more about this interesting phenomenon. It was an interesting phenomenon, both from a human and from a social point of view. The ingredients and therefore the flavour and taste of Garsington altered a little when peace came. The C.O.s—Conscientious Objectors—whom Philip and Ottoline had so generously harboured during the war, of course, drifted away. The C.O.s, being pacifists, were, for the reasons which I have explained above in dealing with Massingham's virulence, more quarrelsome and cantankerous than the average man or woman. At week-ends they formed an unquiet, disquieting, turgid sediment beneath the brilliant surface of very important people, the distinguished writers, cabinet ministers, and aristocrats who sat down to breakfast, lunch, and dinner.

Ottoline is almost always described in the setting of Garsington, but she functioned just as characteristically in the large house in Bedford Square. In the 1920s there were three great London hostesses with would-be salons to which the literary gents and ladies were admitted and, if distinguished, welcomed—Lady Colefax, Lady Cunard, and Lady Ottoline Morrell. The social historian of the period could have studied in these salons the antics of some limited and not uninfluential sections of British society—a way of life,

a collection of human subspecies, and even a form of influence which have, I suppose, completely died out of
London and Britain. The three salons differed a good deal
from one another. Of Lady Cunard's I could only speak
second-hand, from Virginia, for I never went there myself.
Sibyl Colefax was the most professional of the three, an
unabashed hunter of lions.

Ottoline's Bedford Square was even more a salon than
her Garsington. It existed in four forms: you might be
invited to a lunch, a tea, a dinner, or to an evening party
after dinner, and the last might be very large or fairly small.
At all of them the pudding would certainly contain plums,
distinguished or very distinguished persons, and the point of
the pudding was, it seemed to me, not so much in the eating as in the plums—the bigger the better. In the pudding
of society I am not too fond of plums. Nothing is more
enjoyable than 'society', if by the word one means the
gathering together round a table or a fire or in a garden of
congenial, intelligent, and amusing people, and the enjoyment comes from the play and interplay of character and the
congenial, intelligent, and amusing conversation, and is en-
hanced by pleasant or beautiful rooms and houses, good
food, and good wine. This kind of society and its enjoyment
is only possible if the number of people gathered together—
the party—is strictly limited, indeed small enough to make
it possible for the conversation to become at any moment
general. Both Virginia and I were very fond of this kind of
society and party, and we always contrived to get a good
deal of it in Richmond, Rodmell, and later in London.

The society of the professional hostess, of Ottoline in
Bedford Square, is entirely different. As a study of human
behaviour, both of hostess and guests, it always fascinated
me. The psychology of the hostess may contain all or any of
the following ingredients: enjoyment of the enjoyment of

her guests; a kind of artistic creativeness—the art of hostessship; the love of the exercise of power and prestige; the
passion of the collector of anything from stamps to human
beings. The ingredients in the hospitality of Lady Colefax
in Argyll House were quite different from those of Lady
Ottoline Morrell in Bedford Square. Sibyl gave me the impression of an armour-plated, electroplated, or enamelled
woman, physically and mentally. The range of her feelings
behind this metallic façade seemed to be extremely limited;
but façades are façades, and behind hers there may, of course,
have been a tremulous sensitiveness. Indeed I was often
startled and shocked to observe the expression of the eyes in
that mask of her hostess face; far behind and deep down
below they gave one a glimpse of misery, anguish. But the
surface was always hard, polished, plated, professional.
Every morning Sibyl wrote her illegible notes or sat at her
telephone collecting men and women, ranked solely for their
fame or footing, their power or prestige. Her main motives
were, I think, pleasure in power and prestige and the delights
of collecting—'I must add Walter Lippmann and André Gide
to my collection'.

The hostess psychology of Ottoline was quite different. I
do not think that she had a very strong passion for collecting,
although, as with all professional hostesses, it did exist in
her. She was, too, not very much moved by power and prestige; as a Cavendish-Bentinck and sister of the Duke of
Portland, she assumed unconsciously, like all aristocrats, that
she had a peculiar right and relation to both, and therefore
need not trouble about them. She was highly sexed and got
some sexual satisfaction as a by-product of the art of hostessship. She also got aesthetic satisfaction from the practice
of the art, for her aesthetic emotions were strong and persistent, if erratic and sometimes deplorable. The house and
garden at Garsington were lovely, and Ottoline gave both

an artistic finish, and she gave the same to the rooms in Bedford Square. Her own taste was for disorderly flamboyance, as her dress and hair showed, but she knew and respected what the world and the élite thought to be the right thing in books, pictures, music, houses, rooms, furniture, and persons. The compromise between good taste and her own tastes gave a peculiar and sometimes incongruous aspect to her rooms and a strange and sometimes ludicrous flavour to her parties. Her reactions to what is great in art were strong, untrustworthy, and embarrassing; for instance, I have heard her gush over the beauties of Keats at the breakfast table of a Garsington week-end to five or six silent, gloomy, cynical, sophisticated members of the literary élite. But she had a real gift for and pleasure in the art of hostessship which was unknown to Sibyl. She wanted to know, to have intimate relations with intelligent, imaginative, creative people, and to create herself the best possible surroundings in which these strange men (with an occasional woman) might flourish socially and enjoy one another's society and conversation. There is no doubt that in this she was to some extent successful.

If you want to know what a particular period was like, the nature of its society and classes, the kind of people who lived in it, you can learn something from the way in which the people met and entertained one another formally. The *Symposium* gives one a vivid and startling glimpse, not only of Athens in the fifth century, but of Socrates, Alcibiades, and Aristophanes, just as Petronius makes one suddenly see through Trimalchio's dinner-party a glimpse of what it meant to be a vulgar rich man or a slave girl in the time of Nero. Most of those whom I met in Ottoline's Garsington and Bedford Square or Sibyl's Argyll House are as dead as Socrates and Trimalchio, and the society of the 1920s is almost as dead as that of Athens in the fifth century B.C. or of Rome in the first century A.D. If I describe one or two parties at Sibyl's and Ottoline's, it may give a glimpse of what they and we and a section of London society were like in the third decade of the twentieth century.

First a trivial picture of Sibyl, the insensitive professional hostess, and the failure of her art. In *Beginning Again* (p. 167) I described how, just before the 1914 war, I met for the first time the famous Walter Lippmann, then unknown, how we travelled down from Keswick to London talking intimately the whole time, and how much I liked him. Not long before the 1939 war Sibyl came to see us and for some reason which I have forgotten I must have mentioned this. I also said that in the intervening 25 years I had hardly seen Lippmann. Lady Colefax, the pro, jumped on me. Walter Lippmann would be in London next week; would I come and dine and meet him on Thursday? I knew that at a dinner in Argyll House I should have no chance of the only kind of conversation which I wanted to have with Lippmann, and I therefore refused. But Sibyl, as a hostess, was a ruthless Lady Bountiful, and I was not allowed to get off. She would get Lippmann to come and meet me at six one evening, if I would not come to a meal, and she would ask no one else. I foolishly agreed. We met unhappily and, under the inhibiting eye of Sibyl, had nothing to say to each other.

The second picture was at a top-notch, grand evening party at Argyll House. It was a fine, warm summer evening; the large room was full of Sibyl's top-notch lions, political and literary mainly, together with a sprinkling of lesser lights and the stage army of well-fed and well-dressed men and women whose only distinction was that they were habitually asked to this kind of London party. The doors which led from the large room into the garden were open, and the guests strolled about the garden, which was lit by garlands of fairy lamps. There is a certain beauty in this kind of scene, en-

hanced by the fact that among the strollers under the fairy
lamps are the Prime Minister and half the Cabinet, Mary
Pritchard, Margot Asquith, the editor of *The Times*, Max
Beerbohm, and Augustus John. Everything seemed to be
going as it should, when suddenly there came a social catastrophe of the kind which often happened in Argyll House.
We were all summoned into the large room and seated down
to hear a recital by a distinguished French pianist. She was
led up to the piano by Sibyl and began to play. She had played
only a few bars when two or three people came in from the
garden talking loudly, obviously unaware of what was happening in the room. The pianist crashed her hands on the
notes, got up, and walked to the end of the room, where she
sat down, saying in a loud voice and a thick foreign accent:
'I do not play to accompany people talking'. It is a queer
sensation to sit in such a company of 100 to 150 persons,
in full evening-dress, all in awkward silence and all obviously
feeling rather uncomfortable. No one moved, no one talked.
After what seemed a long time Sibyl got up and walked over
to where Arthur Balfour was sitting and had a longish conversation with him. Then he got up, went to the irate pianist,
bent over her, and obviously pleaded with her. He was
successful and led her gracefully amid applause to the piano.

Ottoline, as I said, treated her lions differently, and the
atmosphere of her Bedford Square zoo was much more ramshackle and informal than that at Argyll House. I can,
perhaps, best give its flavour by describing a tea-party there.
I did not often go to these parties, and it was characteristic of
Ottoline that she insisted that I should come to this one,
because, she said, one of the Georgian poets whom I had
never met was coming and she was quite sure I would like
him. His name, I think, was Ralph Hodgson and Ottoline
thought I would like him because he was a strong silent
man who had written a poem about a bulldog and also a

poem about a bull. When Virginia and I arrived, the poet of the bull and the bulldog was there, strong and silent, together with Yeats and James Stephens, who had written a very successful book, *The Crock of Gold*. It was an uneasy party. Yeats sat in the place of honour, but was grumpy and silent, and Virginia was commandeered, much too obviously, by Ottoline to go over and sit next to him and talk him, if possible, into a better mood. James Stephens was one of those Irish Irishmen, the stage Irishman who never stops talking with the soft brogue which makes one think despairingly of the indomitable soft rain falling upon the lakes of Killarney. Being also what I call a literary gent, he used to fill me alternately with depression and irritation, and I think that he probably had much the same effect upon Yeats. On this occasion he was in full spate, with a whimsical, poeticised fantasy about insects, whom he continually referred to as 'the little craytures'. When he made a more than usually absurd statement about 'the little craytures', before I could stop myself, I said in a loud voice: 'Nonsense'. Ottoline frowned upon me and the party became still stickier. However, nothing stopped Stephens talking, and the other Irishman, the great man, thawed bit by bit under the skilful and soothing ministrations of Ottoline and Virginia. The party began to go in the way in which Ottoline liked parties to go—intimate, intense, and rather intensive talk about books and writers and the arts generally. It was this kind of conversation which made Bedford Square so different from Argyll House. And it was characteristic of a Bedford Square party that it was suddenly deflated, broken up, exploded. For the door opened and in came the Duchess of Portland. She was obviously not expected by Ottoline and she looked upon us all as if we were 'the little craytures'. She sat down on a sofa next to Ottoline and began to talk to her about something which only concerned herself. Silence fell upon

us little craytures; even Stephens was left without a drone or
a buzz, for a Duchess of Portland is capable of silencing the
voice even of the cicada. After a minute or two Ottoline got
up and took her sister-in-law out of the room. They stood
outside the door and the sound of their voices in inaudible
conversation seemed to be going on interminably. When at
last we heard the front door close behind the Duchess, the
party rather despondently broke up.

The kind of party which I have just been describing,
presided over by a professional hostess, is formal, public
entertainment in which social pleasure is very deliberately
offered and pursued. I have never found that kind of pleasure
very pleasurable. On the other hand, as I have said, both
Virginia and I enjoyed society, if private, informal, intimate.
As Virginia's health improved and civilization began to
penetrate to Rodmell in the form of a bus and other ameni-
ties, we became more and more social in that kind of way, and
the number of our friends and acquaintances grew rapidly.
We saw them at Richmond, but we also had them for week-
ends to Monks House. Our week-ends at Monks House
were the antithesis of week-ends at Garsington. We had
room for only one guest, and it was still pretty primitive and
uncomfortable, so that one could only have those with whom
one was already very intimate or with whom one could soon
become very intimate. Among the former were Lytton
Strachey and Morgan Forster, and among the latter T. S.
Eliot.

In the years 1920 to 1923 Tom Eliot stayed with us
several times in Rodmell and he used to come and dine with
us at Richmond. It was in these years that our relations
with him changed deeply, from extreme formality to the
beginning of a real intimacy. But I do not think it was
merely that we got to know him better. I think that Tom
himself changed inside himself to some extent—to the extent

perhaps that anyone ever can change inside himself after his first good cry on leaving his mother's womb and seeing the cruel light of day and the face of nurse and doctor. There was from the first a dichotomy in Tom, which, when he stayed with us on September 19, 1920, Virginia noted in his face; 'The odd thing about Eliot,' she wrote, 'is that his eyes are lively and youthful when the cast of his face and the shape of his sentences are formal and even heavy. Rather like a sculpted face—no upper lip; formidable, powerful; pale. Then those hazel eyes, seeming to escape from the rest of him.' He was so inhibited, those sentences were so formal and heavy that, although—or rather because—I had seen so much in his poetry and in those eyes which seemed to escape from him, the week-end left me with a feeling of disappointment. In conversation it was his brain that was disappointing, so much more rigid and less powerful than I had expected from the poems, and with so little play of mind. He was himself aware of this and disappointed in himself, for, in describing a week-end with Ottoline in Garsington, he said: 'And I behaved like a priggish pompous little ass'. I do not think that it was just conceit that made us think that we had something to do with changing Tom, with loosening up the pomposity and priggishness which constricted him, with thawing out the essential warmth of his nature which, when we first knew him, seemed to be enclosed in an envelope of frozen formality. How inhibited he was then can be seen from an absurd incident which happened at one of his very early visits to Monks House and in which I remember for the first time breaking the ice. He was walking with Virginia and me across the fields down to the river. I suddenly wanted to make water and fell behind to do so. Neither of my companions saw what I was doing, but I suppose it was very obvious what I was doing. Anyhow, when I caught them up again, I felt that Tom was uncomfortable, even shocked.

I asked him whether he was and he said yes, and we then had what gradually became a perfectly frank conversation about conventions and formality. Tom said that he not only could not possibly have done what I did, that he would never dream of shaving in the presence even of his wife.

About literature, even about his own writing, even in those early days of knowing him, he was easy and unreticent —and always very interesting. During this visit Virginia one evening tackled him about his poetry and told him that 'he wilfully concealed his transitions'. He admitted this, but said that it was unnecessary to explain; explanation diluted facts. He intended to write a verse play in which the four characters of Sweeney would appear. What he wanted to do was to 'disturb externals'; he had had a kind of personal upheaval after writing *Prufrock*, and this altered his inclinations, which had been to 'develop in the manner of Henry James'.[1]

Tom had a great opinion of Virginia as a critic. Some ten years after this early visit to Rodmell, he came to us one day and said that he had just written some poetry which he would very much like us to criticize seriously. What he would like would be to send us each a typescript of the poems; we should read them and then come in after dinner one evening and each in turn criticize them—and he might ask one or two other people—Mary Hutchinson, for instance—to come as well. We agreed, and he sent us a typescript of what eventually was published as *Ash-Wednesday*. Then one summer evening we went round to his house after dinner and found Mary Hutchinson and McKnight Kauffer there. We all sat solemnly on chairs round the room and Tom began the proceedings by reading the poem aloud in that curious monotonous sing-song in which all poets from Homer downwards have recited their poetry. Then each in turn was called upon

[1] Some of the above is recorded in Virginia's diary.

to criticize. The order was, I think, Mary, I, Virginia, Kauffer. It was rather like an examination, not of the examinee, but of the examiners, and Mary, Kauffer, and I didn't do any too well—in fact Tom dismissed rather severely some of the things that some of us said. Virginia passed with flying colours. She told Tom that he had got into the habit of ending lines with a present participle; he had done it with great effect at the beginning of *The Waste Land*, and he was doing it again in this poem. She thought he should beware of it becoming a habit. Tom said that she was quite right and that what she said was very useful. I still have the original typescript copy which he gave to me and I have compared it with *Ash-Wednesday* as published. The two versions are not the same, but the present participles remain in the lines. The printed version is:

> *Here are the years that walk between, bearing*
> *Away the fiddles and the flutes, restoring*
> *One who moves in the time between sleep and*
> *waking, wearing . . .*

In the original typescript the three lines read:

> *Here are the years that walk between, bearing*
> *Away the fiddles and the flutes, restoring*
> *One who walks between season and season, wearing . . .*

In the book as published, there are six sections, in the typescript there are only five; presumably section VI was written after our critical seance. Tom did not alter very much in the first five sections; there are two main differences between the two versions. In the typescript each section has a title which is not in the printed book and the final version of section V is much longer than the original. The following are the titles of the sections:

PEACE IN OUR TIME, O LORD

I. PERCH'IO NON SPERO; II. JAUSEN LO JORN; III. SOM DE L'ESCALINA; IV. VESTITA DI COLOR DI FIAMMA; V. LA SUA VOLUNTADE.

The gradual growth of our intimacy with Tom can be traced in Virginia's diary. In February 1921 he dined with us at The Cock in Fleet Street and Virginia wrote: 'pale, marmoreal Eliot was there last week, like a chapped office boy on a high stool, with a cold in his head, until he warms a little, which he did. We walked back along the Strand. "The critics say I am learned and cold," he said. "The truth is I am neither." As he said this, I think coldness at least must be a sore point with him.' A month later she was wondering whether we would ever get to the stage of Christian names: 'But what about Eliot? Will he become Tom? What happens with friendships undertaken at the age of 40? Do they flourish and live long? I suppose a good mind endures, and one is drawn to it and sticks to it, owing to having a good mind myself. Not that Tom endures my writing, damn him.' By the end of the year we were calling him Tom and Virginia noted with regret that she was no longer frightened of him.

The first time Virginia met Vita Sackville-West (Mrs Harold Nicolson) was in December 1922, and the first entry in her diary describing Vita is rather critical. We saw something of Harold and her during the next year, but it was not until 1924 that we got to know them well. At that time they lived partly in London and partly in a very pleasant house, Long Barn, near Sevenoaks, and not far from her ancestral home, Knole. We stayed with them there, and Virginia began to see a great deal of Vita. There was a curious and very attractive contradiction in Vita's character. She was then literally—and so few people ever are literally—in the prime of life, an animal at the height of its powers, a beautiful flower in full bloom. She was very handsome, dashing, aristocratic, lordly, almost arrogant. In novels people often 'stride' in or out of rooms; until I saw Vita, I was inclined to think that they did this only in the unreal, romantic drawing-rooms of the novelist—but Vita really did stride or seem to stride.

To be driven by Vita on a summer's afternoon at the height of the season through the London traffic—she was a very good, but rather flamboyant driver—and to hear her put an aggressive taxi driver in his place, even when she was in the wrong, made one recognize a note in her voice that Sackvilles and Buckhursts were using to serfs in Kent 600 years ago, or even in Normandy 300 years before that. She belonged indeed to a world which was completely different from ours, and the long line of Sackvilles, Dorsets, De La Warrs, and Knole with its 365 rooms had put into her mind and heart an ingredient which was alien to us and at first made intimacy difficult.

Vita was, as we used to say to her, only really comfortable in a castle, whereas a castle is almost the only place in which I could not under any circumstances be comfortable. When compared to the ramshackle informality of our life and rooms in Hogarth House and Monks House, Vita's Long Barn, with its butler, silver, Persian rugs, Italian cabinets, and all other modern conveniences, seemed to us a house and a way of life of opulence and grandeur. In their own way—which happens not to be my way[1]—both the house and the way of life had considerable charm and beauty. Later Vita's passion

[1] Virginia, on the whole, liked rather more than I did the conventional opulence of the life and habitations of the wealthy English upper classes. But returning from a week-end at Long Barn, after describing the 'opulence', she added: 'Yet I like this room better perhaps; more effort and life in it, to my mind, unless this is the prejudice one has naturally in favour of the display of one's own character'.

for castles led her to buy the great tower and ruined buildings of Sissinghurst. As the thousands of people who every year visit Sissinghurst know, she restored a good deal of the castle and created a garden of very great beauty.

In the creation of Sissinghurst and its garden she was, I think, one of the happiest people I have ever known, for she loved them and they gave her complete satisfaction in the long years between middle age and death in which for so many people when they look out of the windows there is only darkness and desire fails. But there was another facet in her character; she was, in many ways, a very simple person, and it was this side of her which emerged both in her poem, *The Land,* and in her passion for gardening, though combined with the opulent magnificence of the Sackvilles and Knole it produced something at Sissinghurst which could not exactly be called simple. It was this simplicity, when combined with other things in her character, which made one fond of her, the other things her own affectionate nature and her honesty and generosity. The scale of Sackville generosity in those days was to some extent, I think, influenced by the crazy munificence of Vita's mother (described in her book *Pepita*) who in her life had dissipated several million pounds without leaving herself anything to show for it. Lady Sackville lived eccentrically at Rottingdean and Vita used from time to time to drive over to see her. On the way back she used to look in upon us at Monks House and we always went out to her car to see the presents which Lady Sackville had showered on her. The scale of her munificence can be seen in the fact that one day on the back seat of the car was a gigantic porcelain sink in which were piled about 150 green figs. I don't know why Lady Sackville had given Vita a porcelain sink, but the figs came about in this way. She took Vita for a drive to the famous fig garden in Worthing; she asked Vita whether she would like to take back some figs, and, when she said yes, insisted upon buying for her the entire crop of ripe figs.

I find some difficulty in determining exactly when what is called Bloomsbury came into existence. In *Beginning Again* (pp. 21-26) I treated it as having come into existence in the three years 1912 to 1914. I should now prefer to say that in those three years a kind of ur-Bloomsbury came into existence. Of the thirteen members of Old Bloomsbury, as we came to call it, only eight at that time actually lived in Bloomsbury: Clive and Vanessa in Gordon Square and Virginia, Adrian, Duncan Grant, Maynard Keynes, and myself in Brunswick Square, with Saxon Sydney Turner in Great Ormond Street. It was not until Lytton Strachey, Roger Fry, and Morgan Forster came into the locality, so that we were all continually meeting one another, that our society became complete, and that did not happen until some years after the war. First the war scattered us completely and then Virginia's illness, by banishing us to the outer suburb of Richmond, made any return to our day-to-day intimacy impossible. But as Virginia's health improved and it became possible for us to go up to London more often to parties and other meetings, what archaeologists might call a second period of ur-Bloomsbury began. For instance, in March 1920 we started the Memoir Club and on March 6 we met in Gordon Square, dined together, and listened to or read our memoirs.

The original thirteen members of the Memoir Club, identical with the original thirteen members of old Bloomsbury, were all intimate friends, and it was agreed that we should be absolutely frank in what we wrote and read. Absolute frankness, even among the most intimate, tends to be relative frankness; I think that in our reminiscences what we said was absolutely true, but absolute truth was sometimes filtered through some discretion and reticence. At first the memoirs were fairly short; at the first meeting

seven people read. But as time went on, what people read became longer and, in a sense, more serious, so that after a few
years normally only two memoirs were read in an evening. They
were usually very amusing, but they were sometimes something more. Two by Maynard were as brilliant and highly
polished as anything he wrote—one describing his negotiations with the German delegates and, in particular, Dr
Melchior, in the railway carriage at Trèves after the 1914 war,
and the other about Moore's influence upon us and our early
beliefs at Cambridge—and these were after his death published, exactly as they were originally read to us, under the title *Two
Memoirs*. Some of Virginia's were also brilliant, and Vanessa
developed a remarkable talent in a fantastic narrative of
a labyrinthine domestic crisis. The years went by and the Club
changed as the old inhabitants died and the younger generation
were elected. The last meeting took place, I think, in 1956, 36
years after the first meeting. Only four of the original thirteen
members were left, though in all ten members came to the
meeting.

These meetings meant for us going up to Bloomsbury
from Richmond, with late nights, staying in London or midnight train journeys back to Richmond. And we were sucked
into other parties both in and outside Bloomsbury. In order
to give a more concrete idea of one of these parties in ur-
Bloomsbury and of Virginia's social excitement which I
have referred to when she found herself at one, I will quote
from Virginia's diary the description of a fancy-dress party
to which we went in Gordon Square in the first week of
January 1923:

Let the scene open on the doorstep of number 50
Gordon Square. We went up last night, carrying our bags
and a Cingalese sword. There was Mary H. in lemon
coloured trousers with green ribbons, and so we sat down

to dinner; off cold chicken. In came Roger and Adrian and Karin; and very slowly we coloured our faces and made ready for number 46. It was the proudest moment of Clive's life when he led Mary on one arm and Virginia on the other into the drawingroom, which was full, miscellaneous, and oriental for the most part. Suppose one's normal pulse to be 70: in five minutes it was 120: and the blood, not the sticky whitish fluid of daytime but brilliant and prickly like champagne. This was my state and most people's. We collided, when we met; went pop, used Christian names, flattered, praised, and thought (or I did) of Shakespeare. At any rate I thought of him when the singing was doing. Shakespeare I thought would have liked us all tonight....My luck was in though and I found good quarters with Frankie [Francis Birrell] and Sheppard and Bunny [David Garnett] and Lydia [Lopokova]—all my friends in short. But what we talked about I hardly know. Bunny asked me to be his child's godmother. And a Belgian wants to translate me. Arnold Bennett thinks me wonderful and ... and ... (these, no doubt, were elements in my hilarity). Jumbo [Marjorie Strachey] distorted nursery rhymes: Lydia danced: there were charades: Sickert acted Hamlet. We were all easy and gifted and friendly and like good children rewarded by having the capacity for enjoying ourselves thus. Could our fathers? I wearing my mother's laces, looked at X's Jerboa face in the old looking glass—and wondered, I daresay no one said anything very brilliant. I sat by Sickert and liked him, talking, in his very workmanlike but not at all society manner, of printing and Whistler; of an operation he saw at Dieppe. But can life be worth so much pain, he asked. 'Pour respirer,' said the doctor. 'That is everything.' 'But for two years "after my wife's death" I did not wish to live,' said Sickert. There is something indescribably congenial to me in this easy artists' talk: the values the same as my own and therefore right: no impediments: life charming, good and interesting: no effort: art brooding calmly over it all: and none of this attachment to mundane things, which I find in Chelsea. For Sickert said, why should one be attached to one's body and breakfast? Why not be satisfied to let others have the use of one's life and live it over again, being dead one-self? No mysticism, and therefore a great relish for the actual things—whatever they may be—old plays, girls, boys, Proust, Handel sung by Oliver [Strachey], the turn of a head and so on. As parties do, this one began to dwindle, until a few persistent talkers were left by them-selves sitting in such odd positions. . . . And so, at 3, I suppose, back to No. 50 to which Clive had gone pre-viously.

It was parties like this and our increasing sociability in London which made the question of whether we should stay on in Richmond or immigrate to Bloomsbury more and more urgent. Already in 1922 Virginia was eager for the move. She had begun to feel imprisoned, secluded and excluded, in Richmond. If she lived in London, she said, 'I might go and hear a tune, or have a look at a picture, or find out some-thing at the British Museum, or go adventuring among human beings. Sometimes I should merely walk down Cheapside. But now I'm tied, imprisoned, inhibited.' This was, of course, true, but I had been against a move, solely because I feared the result on Virginia's health. It had be-come much more stable, but it could never be neglected or ignored, and nothing was more dangerous for it than the mental fatigue produced by society and its social pleasures. She was one of those people who drained herself, exhausted herself mentally, both passively and actively, not only at a party but in any kind of conversation or social intercourse.

In Richmond it was possible to keep some control over our social life and, at a danger signal, to shut ourselves off from it for a time. I feared that this would prove impossible in London.

However, in the middle of 1923 I became converted to the idea, for the disadvantages of staying on in Richmond seemed to outweigh the dangers of moving to Bloomsbury. I had to count the cost both of Virginia's growing feeling of being cabined and confined and, as our engagements in London increased, the increasing strain and fatigue of catching crowded trains or buses in order to keep them. So I gave way and in November we began to search Bloomsbury for a house. The usual alternation of joy and disappointment and despair in house hunting lasted for two months, but on January 9, 1924, we acquired from the Bedford Estate a ten-year lease of 52 Tavistock Square. On March 13 following, we moved into it.

Chapter Two

DOWNHILL TO HITLER

52 Tavistock Square, eventually destroyed in 1940 by one of Hitler's earliest bombs, was a very pleasant house, built, I suppose, by a Duke of Bedford, as speculative builder, early in the nineteenth century. It had four storeys and a basement. We put the Hogarth Press into the basement and ourselves occupied the flat on the second and third floors; the ground and first floors were already let to a firm of solicitors, Dollman and Pritchard. The firm actually consisted of old Mr Pritchard and his son, young Mr George, and a strange staff. First there was old Mr Pritchard's sister, who had, I believe, been matron in a large London hospital but now acted with immense efficiency as a kind of head clerk. Then sitting by himself in a room on the first floor was a most sophisticated Irishman, who had lived a good deal of his life in Paris and spoke perfect French. Finally there were two or three girl clerks. One would have to live many lives and travel a long way to find again as good tenants as the firm of Dollman & Pritchard; we got so friendly with the partners and all the staff that when we moved in 1939 from Tavistock to Mecklenburgh Square, we took the firm with us. Old Mr Pritchard, in looks, in speech, and in character, came straight out of Dickens, belonging to the year 1850 rather than 1924 and to the long line of angelic old men in the great Victorian's novels. The Pritchards were so absurdly generous that, when I employed them as solicitors, I found the greatest difficulty in persuading them to charge me anything.

I have lived practically all my life in London, but I am again and again surprised by its curious, contradictory character, its huge, anonymous, metropolitan size and its pockets of provincial, almost village life—also the congenital conservatism of Londoners, so that if you scratch the surface of their lives in 1924 you find yourself straight back in 1850 or even 1750 and 1650. In the 15 years I lived in Tavistock Square I got to know a gallery of London characters who themselves lived in a kind of timeless London and in a society as different from that of Fleet Street, Westminster, Kensington, or Putney—all of which I have known—as Sir Thomas Bertram's in Mansfield Park must have been from Agamemnon's behind the Lion Gate in Mycenae. There was still in it a strong element of Dickensian London. There was a perfect Dickens character not only at the top, but also at the bottom of Dollman & Pritchard, solicitors. When at 5 or 6 o'clock the business closed its doors and their rooms were empty, the front door was pushed open and in sidled Mrs Giles. She wore, as Kot would have said, a haggish look; indeed, she always reminded me of one of those wizened, skinny, downtrodden, grimy women of the Dickensian underworld. She was the Platonic image of the London char laid up, not in heaven, but in an attic in Marchmont Street, London, W.C. 1. She spoke the language of the Dickensian cockney. And when she was in a room by herself, she talked as, I am quite sure, Dickens's gaunt old women spoke when they were outside his novels—a language which would have been quite impossible in a Victorian novel. It was terrifying. When just before dinner I used to take my dog out into the Square garden for a run, as I passed through the hall, I used to hear Mrs Giles talking to herself as she cleaned Mr Pritchard's front room. It was a monotonous stream of the foulest language which I have ever heard. Some of it was just pure, disinterested swearing for the sake of swearing, but every now and again it turned into a particularized hymn of hate, the most horrible, obscene accusations against the various members of the firm. I cannot attempt to explain this strange phenomenon. There was little or no bitterness in Mrs Giles's voice—it was not so much vituperation as a monotonous threnody, a lament for the horrors of Mrs Giles's life—and all the Mrs Giles's lives—in Marchmont Street.

I used to take my dog for a run in the Square three or four times a day and I therefore walked in it far more often than any other resident, and I soon got to know the Square keeper very well. After a year or two I was elected to the Square Committee. We were a statutory body, and here again I found myself back in the London of Dickens. The use and government of Tavistock Square were regulated by by-laws which dated, I think, from about 1840. The Square Committee, which administered the by-laws, consisted of three persons annually elected by householders resident in the Square and each householder was entitled to a key admitting him into the garden. One by-law laid it down that no man-servant or maidservant should be allowed in the Square and another prescribed what games children should be permitted to play there. The Square keeper was another type of nine-teenth-century cockney. The square was for him the centre of a large village, bounded on the north by the Euston Road, on the east by the Grays Inn Road, on the south by Russell Square, and on the west by Tottenham Court Road. Within those boundaries few things happened, at any rate of a discreditable nature, which he did not learn, and, when I got to know him well, which he did not recount to me at great length. He knew by sight almost all those whom he counted to be true inhabitants of this Bloomsbury village and he had an extraordinary knowledge of the private lives of very many of them. Many of his stories were libellous and most of them were, I think, true. He was a sardonic, poker-faced, dis-illusioned man who noted and described, without heat or any sign of moral indignation, the frailties and rascalities of human nature in the rich, the poor, and the police between Tottenham Court Road and Grays Inn Road.

The police were not popular either with my friend or with the poor in Bloomsbury. Their chief victims, according to the Square keeper, were prostitutes, barrow-boys, and eating-house keepers. There was a regular tariff which the first two paid in order to pursue their business in peace. It was almost impossible for restaurant keepers not to break the law occasionally, and many of them insured against the consequences by providing free meals for any policeman from whom trouble might be expected.

I had one absurd brush with a policeman which at the time seemed to me to confirm my friend's view of the force. One Sunday night about 11 Virginia and I were walking down Francis Street towards Tavistock Square, returning from an evening with Vanessa in Fitzroy Street. Towards us came a large woman about 30 to 35 years old, rather drunk and staggery. Some way behind her also walking towards us came a policeman. Two men on the opposite side of the road began to jeer at her. She stopped and let loose a volley of abuse about 'bullocks' balls' across the road at them. They replied, but, catching sight of the approaching policeman, bolted down a side street, dropping in their haste a bottle of beer with a crash on the pavement. The policeman came up to the woman and began to hector her. She was in no yielding mood, and it seemed to me that he was deliberately trying to goad her into doing something which would justify an arrest for being drunk and disorderly. I suddenly lost my temper and dashed in between them, telling her to stop talking and to leave things to me. I told the policeman that he must have seen that it was the two men who had started the whole thing—why had he done nothing to them and begun to hector the woman? An argument began and I suddenly realized that we were already surrounded by a small crowd which murmured its support of me and the woman against the policeman. I gave my name and address to the policeman and told him that, if he prosecuted the woman, he would have to call me as a witness. The woman, finding that she was supported, then turned on the policeman and began to abuse him. The delighted crowd increased and things began to look rather unpleasant, so I turned to a rather sensible looking man who had been standing by my side and asked him to take the woman away before she got into more trouble. He and I induced her to go and I then had once more to face the policeman. Finding that everyone was against him, he was suddenly deflated and apologetic, assuring me that he never meant to charge the woman. We parted almost amicably and, as the crowd broke up, I saw Lydia standing on the outskirts under a gas lamp, gazing with amazement at me and the policeman.

52 Tavistock Square was, as I have said, divided into three parts: the top inhabited by Virginia and me, the middle by Dollman & Pritchard, and the basement by the Hogarth Press. There were four rooms in the basement. The Press used the large room, which had once been the kitchen, as its office, occupied, when we moved in, by our only employee, Marjorie Joad. We printed in what had been the ancient scullery. Then there was a small room at the back, and behind that the old billiard room, which I have described on page 53.

Virginia was such a bad sleeper and so disturbed by noise that, although she habitually used earstoppers, when we first came to Tavistock Square, she thought that, if she slept upstairs in the flat, the noise of the traffic would keep her awake. The first night, therefore, we had her bed put in the small room at the back of the basement and she started to sleep there. But in the middle of the night she was awakened by several rats scampering round her bed on the floor, and she had to retreat up to the flat for good.

The rats were the harbingers of much trouble and of a legal case which I had to bring in the High Court of Justice. The rats came from a large open space, full of old bricks and rubble, which fronted Woburn Place and stretched from the back of Tavistock Square to Russell Square. The Imperial Hotel Company had acquired this site and were clearing it in preparation for building what was eventually the Royal Hotel. The result was a vast number of displaced refugee rats who could be seen in the middle of the day searching the dustbins in the area and who invaded the basement of nights. With the help of a London County Council inspector who looked after vermin, we got rid of the rats, but the building of the hotel caused us a great deal of trouble. While the building operations were actually going on, the noise during the day was pretty bad, and we had double windows put into our main sitting-room, which looked straight into the area of devastation and the scaffolding, in order to keep out the creaking of the cranes, the clanking of lorries, the cries and curses of the builders.

But our troubles really began only after the vast hotel was finished. At the back was a long ballroom the windows of which were immediately below the windows of our sitting-room. When in the evening these windows were open and a jazz band was playing full blast from 8 to 12, the Bedlam of noise, funnelled into our room even with the double windows closed, made life impossible. I wrote to the Imperial Hotel Company complaining, and got back a sympathetic letter from the secretary, promising steps to mitigate the noise. The only possible mitigation was by closing the hotel windows and I think they gave orders that this should be done, but even if their dance began with closed windows, someone always opened them half-way through the evening, letting Bedlam loose into our sitting-room. I went round to see the secretary, who was very friendly, especially when he found that I had known his famous brother, E. C. Bentley, the author of *Trent's Last Case*. (E. C. Bentley and G. K. Chesterton had been members of a debating society to which I belonged, when a boy at St Paul's, as I related in *Sowing*, p. 100.) But in the end I had to take legal action, which turned out to be an interesting experience.

I employed Mr Pritchard as my solicitor. We eventually won our case, and it is, I think, the only serious case which, in my experience, has been taken to court, has been won, and has cost the plaintiff not one single halfpenny—we recovered all our costs. But the proceedings made me realize, as I had not before, the precariousness and helplessness of the individual in modern life. It was essential to prove that the noise was legally 'a nuisance', a nuisance which anyone or everyone would find intolerable, not just a noise which some hypersensitive person would object to. So I went round and canvassed all the residents whose rooms in Tavistock Square and Upper Bedford Place looked on to the back of the hotel, to get them to come and give evidence that the noise was a nuisance to them. Every man and woman whose windows were near the hotel windows said that the noise made life intolerable to them in the evenings when there was a dance band playing, and that they would give evidence, but in the end, when it came to the point, not one single person would do so—they all cried off. The reason was that they all had short leases and were frightened of their landlord. The rumour was that their landlord, the Bedford Estate, was in some way interested in the hotel operations and was going to sell more property to the hotel company, and that, if any one took action against the company, he would probably find that he could not get a renewal of his lease.

Whether there was any ground for the rumour or for their fears, I do not know. The important point was that these people believed the rumour and, because of their fear, were prepared without resistance to allow 'them'—an impersonal company and an impersonal landlord—to make their lives a burden to them. The tyranny of these impersonal or personal 'thems' has of course always been a terrifying menace hanging over the lives of ordinary people. 'They' used to be kings, aristocrats, classes, and churches, and it was thought at one time that liberty, equality, and fraternity would abolish them, but 'they' have continued to exist and flourish under other names. In 1965 in a Sussex village a man, with a wife and three children, who has worked all his life until the age of 40 on a farm in the village in which he was born, becomes ill and unable to do heavy farm work. He receives legal notice to leave the house, which is required for the man who is to take his place on the farm. There is no other house available for him, but the Rural District Council will find him lodgings, separating him from his wife and family, who will be 'accommodated temporarily in an institution'. In the country, of course, the working classes have always lived in tied cottages and, completely in the power of 'them', in the shadow of fate and eviction. But here in Tavistock Square in the 1920s were middle-class people living under the same kind of menace.

Despite the fact that it became clear that no one would give evidence on my side, I decided to go it alone, for I knew that I had a very strong, if not impregnable, case. For I had up my sleeve a sentence in one of the Company's letters to me which in fact admitted that the noise was legally a nuisance. I was convinced that, when they were confronted in court with that letter, they would be unable to plead that the noise was not a nuisance, and that the only question would be what steps we could force them to take in order to 'abate' the nuisance. We were encouraged by the fact that, when the case was put down for hearing, they more than once applied for an adjournment. Eventually they came to us and asked us to agree to a settlement out of court. After some haggling it was settled that judgment should be entered against them, the terms being that the offending windows would always be screwed up if a band was playing and that all our costs would be paid by the Company. Mr Pritchard managed to get from them every penny of our costs. On the whole the nuisance was satisfactorily abated, though every now and again someone would forget to screw up the windows and our room and ears would be filled with din of jazz music. An irate telephone message from me to the hotel would then be required in order to get the nuisance once more abated.

Throughout my life I have always said to myself—and often to other people—that one should change one's occupation every seven years. The first person to discover this important truth appears to have been an ancestor of mine some thousands of years ago, for it is recorded in the 29th chapter of the book of Genesis that Jacob agreed to serve Laban for seven years for his daughter Rachel, and when Laban swindled him with Leah instead of Rachel, he agreed to work another seven years for Rachel. The number seven has, of course, for ages had some mystical attraction for human beings, connected perhaps with the seven days of the week and the seven stars of the Pleiades. It is possible that the mystical nature of the number had something to do with Jacob's offer to serve seven years for a wife and twice seven years for two, though his fiddling with the ringstraked and speckled goats, so practical and ingenious, showed that he was a pretty tough and hardheaded man. I do not think that I am much influenced by the mysticism of numbers or anything else. My feeling about what should be the length of time of an occupation is based on observation of myself and other people: no matter how interesting and complicated a profession or business or occupation may be, after about five years 90 per cent. of it tends to become stereotyped and automatic, and after another two years 99 per cent. is performed with skill and efficiency, but as routine and habit and with about as much thought and originality as the spider has in the last 700,000 years given to its occupation of spinning, with great skill and efficiency, its web. And when that happens, it is time, I (unlike the spider) think, to make a change.

Throughout my life, to quite a considerable extent, I have taken my own advice and septennially changed my occupation. I began with the Ceylon Civil Service in 1904 and resigned from it in 1911 after seven years. From 1915 to 1923—about eight years—I earned a living mainly by journalism, writing for the *New Statesman*, editing the *International Review*, on the staff of the *Nation*. When the *Nation* changed hands in 1923 I was offered and accepted the post of literary editor. At the end of 1930, after rather over seven and a half years, I resigned. As I explained in *Beginning Again* (p. 132) my work as literary editor revealed to me the corroding and eroding effect of journalism upon the human mind, and I have never after 1930 taken a paid job—I have earned my living from the Hogarth Press, from my books, and from occasionally writing articles or reviews.

But for the first seven years in Tavistock Square the diurnal pattern of our lives was in the main drawn for us by my job on the *Nation*. Our offices were in Great James Street, a pleasant building dating from 1720. Every week I spent two or two and a half days in the office, but I also did a considerable amount of work at home, for I wrote a weekly article of about 1,200 words, called 'the World of Books'. For this article I had to read, on an average, two or three books, and I also had to read a rain of articles and poems with which journals like the *Nation* are perpetually deluged. No

Man Ray

Virginia

Monks House, Rodmell

Virginia and Lytton Strachey
By permission of Mrs. Igor Vinogradoff

The author and John Lehmann
at Monks House

Rodmell Village

Gisèle Freund

The author, Sally, and Virginia in Tavistock Square

T. S. Eliot at Monks House

Virginia and Dadie Rylands at Monks House

Mitz in Rome

Mitz at Monks House

Mitz and Pinka

Maynard Keynes and Kingsley Martin
at Rodmell

The author and Nehru
Sport and General Press Agency

Group at Sissinghurst Ben, Virginia, Vita, Nigel

The author at Monks House

one who has not been an editor and/or a publisher can have any idea of how badly how many people can write—and what is even more astonishing than the number and badness of the writers and writings is the belief or even hope that such lamentable stuff could be accepted and printed. It was just the same in the Hogarth Press as it was on the *Nation*: manuscripts poured in upon us and quite a number of them were fatuous and sometimes ludicrously fatuous. Indeed, when they were sufficiently fatuous, they sometimes acquired a quality of such sublime craziness or profound stupidity that we seriously considered starting a 'Hogarth Worst Books of the Year Series' in which we could publish some of them. The number of people who today 'seriously' write books, articles, and poems must be colossal, and I doubt whether one in a hundred thousand of the manuscripts which they produce is ever published. Certainly it was extraordinarily rare to get an unsolicited manuscript which one could accept for the *Nation* or the Hogarth Press. One is inclined to believe that this universal itch of writing is a disease of universal education and the twentieth century—until one remembers that nearly two thousand years ago Juvenal noted it in its Roman form, *cacoethes scribendi*.

As literary editor I was responsible for getting the 'middles', which was the name given to the two or three articles of a literary or general non-political nature, the reviews of books, and regular articles on plays, pictures, music, and science. I doubt whether any weekly paper has ever had such a constellation of stars shining in it as I got for the *Nation*. Here are some of the writers who wrote articles or reviews for me in the first few months of my editorship: Bertrand Russell, G. Lowes Dickinson, Gorky, Augustine Birrell, Roger Fry, E. J. Dent, Walter Sickert, T. S. Eliot, Virginia Woolf, E. M. Forster, Lytton Strachey, Osbert Sitwell, Richard Hughes, Stella Benson, Robert Graves,

V. Sackville-West, Arnold Toynbee. Many of these were in 1923 and 1924 unknown young men and women. In the last year of my editorship most of them were still writing for me and they had been joined by other writers, e.g. Aldous Huxley, L. B. Namier, and Raymond Mortimer.

The literary side of the *Nation* benefited by my running the Hogarth Press, and the Hogarth Press benefited by my being literary editor of the *Nation*. In the Press we were interested in young, unknown writers whose work might not attract the publishing establishment. A journal like the *Nation* puts into one's hand a very wide net and all sorts of literary fishes, large and small, swim into it. All sorts of literary fish, some the same as and some different from the *Nation's* shoals, swam in and out of the Hogarth Press in Tavistock Square. It was possible to help the budding (and sometimes impecunious) Hogarth author by giving him books to review and articles to write; and, if one came across something by a completely unknown writer which seemed to have something in it, one could try him out with articles and reviews before encouraging him to write a book. Thus Tom Eliot, Virginia, E. M. Forster, Robert Graves, Vita Sackville-West, William Plomer came to the *Nation* via the Press, while Edwin Muir was an example of the reverse process in this shuttle service. Somewhere or other I saw a poem of Muir's which I thought very good, so I wrote and asked him to let me see some more. He sent me a few and I published one of them. I got him to come and see me and this began a friendship which lasted to his death. I offered to give him regular reviewing, and from the middle of 1924 for a long time almost every week he had a review in the *Nation*. In 1925 the Hogarth Press published his *First Poems*, in 1926 *Chorus of the Newly Dead*, in 1926 his first book of criticism *Transition*, and in 1927 his novel *The Marionette*. Nearly forty years later we published his autobiography (originally published by Harrap, but considerably enlarged for us) and his last book, *The Estate of Poetry*, was published by us posthumously in 1962.

I look back on this forty years' connection with Edwin Muir with great pleasure and some sadness. We printed his poems in 1925 with our own hands and he was the kind of author and they were the kind of poems for whom and which we wanted the Press to exist. *Chorus of the Newly Dead* was not a book which in 1926 an ordinary publisher would have looked at. We made it a 16-page book, bound in a stiff paper cover, for which we charged 2s. 6d. As we printed and bound the edition of 315 copies ourselves, the actual cost was negligible, i.e. £6, 4s. 7½d. In the first year we sold 215 copies, and after paying the author £3, 18s. 11d., the Hogarth Press had made a profit of £7, 17s. od. Muir was a real, a natural poet; he did not just 'write poetry', the sap of poetry was in his bones and veins, in his heart and brain; that is why, as with practically all real poets, the form and substance of his poetry changed and developed all through his life as he and his mind changed, hammered upon by the grim reality of living. The form and substance of *First Poems* and *Chorus of the Newly Dead* are tremendously different from those of his later poems—so different that, as he tells us in his autobiography, when he reread *Chorus of the Newly Dead* thirty years after he had written the book, the Edwin Muir of 1926 seemed quite strange to the Edwin Muir of 1954. When he reread, he says, the three lines

> *that ghostly eternity*
> *Cut by the bridge where journeys Christ*
> *On endless arcs pacing the sea.*

they seemed 'so strange to me that I almost feel it was someone else who wrote them; yet that someone was myself'. He was an admirable critic. He was so sensitive, intelligent, and honest minded that, as a serious critic, he always had something of his own worth saying even about masterpieces buried long ago under mountains and monuments of criticism. But even in the ephemeral and debased form of criticism, reviewing, he was remarkable. For a long time he used to review novels for me in the *Nation*, a mechanized, mind-destroying occupation for most people. For him it never became mechanical and his mind's eye was as clear and lively after a year of it as when he began.

Edwin's wife, Willa, was also what Koteliansky called a real person and an original writer whose books the Hogarth Press published. They both came from Orkney. An aura of gentleness, soft sea air, the melancholy of remote islands set in turbulent seas surrounded them. All this is too in Edwin's autobiography which we published in 1954. He was the most uncomplaining and unselfpitying of men. I said that I looked back upon my friendship with him with some sadness —the sadness comes from a feeling that life dealt rather hardly with him.

The main interest of my work on the *Nation* came from the people with whom my work brought me into contact. Many of them became my friends. But they were often merely strange adventures, the absurd comedies or tragi-comedies of real life which always astonish and fascinate me. I will give some examples. One afternoon there walked into my room at the *Nation*, Roy Campbell, whose poetry at that time was creating something of a stir. I knew him only slightly. He was dressed in or swathed in one of those great black cloaks which conspirators wear in operas and melodramas and he had a large black sombrero. He sat down, scowled at me, and then said in the peculiar voice which the villain always used in old-fashioned melodramas: 'I want to ask you whether you think I ought to challenge Robert Graves to a duel'. Experience as editor or publisher soon

teaches one that authors, like Habbakuk, are capable of
anything, but I was so astonished by what I heard that I
could only gasp: 'But why?' 'Why?' he said, 'Why? Don't
you remember the review he wrote of my book, the review
you yourself published two weeks ago?' It was true that
Robert had reviewed Campbell's book, but it had never
struck me that there was anything in it to drive the most
hypersensitive writer into lunacy. For the next quarter of
an hour I had a lunatic conversation persuading Campbell
that the laws of honour and chivalry obtaining in Great
James Street in 1926 did not require him to fight Robert
Graves.

Not long after this business I had trouble with another
reviewer, Richard Aldington. I do not remember how I came
across this disgruntled man, who was almost as prickly as
Roy Campbell. He became a regular reviewer for me, the
kind of reviewer who is a godsend to literary editors. He
could and would write me a good review of almost any book
which I sent him—never a very good and never a bad,
always a good review. One day he came to me with a face
almost as gloomy and threatening as Campbell's. He said
that in the last issue of the *Nation* I had had a review by a
Mr X—did I know that Mr X had run off with his (Aldington's) wife? I said that I did not know this and mumbled that
I was sorry to hear it. 'And are you going to employ Mr X as
a reviewer?' said Aldington louringly. I said that Mr X had
written quite a fairly good review and that I would certainly
send him more books from time to time if he continued to
do well. I was immediately presented with a formal ulti-
matum. Unless I gave an undertaking never to employ Mr
X again, Aldington would never again write a review for
me—he could not write in the same paper as the man who
had run off with his wife. I said that I did not think this
reasonable; as a matter of editorial principle, I did not think

it right to give an undertaking to A not to send books to X
to review merely because in a private capacity X had run off
with A's wife. A angrily left me and I do not think that I
ever saw him again.

The strangest of all the incidents which came out of the
Nation was the case of Mr Y. It began one morning when I
was working at home and I received a telephone message
from the office saying that a Mr X had called and wanted
urgently to see me—he could disclose his business only to
me. I told them to tell him that, if he came round to Tavistock Square at once, I would see him for a few moments.
Twenty minutes later there appeared a small gentle-voiced
man in sandals. He told me the following story. He was a
New Zealander and when a youth had become a great friend
of another youth, Mr Y. Mr Y had been a good deal more
affluent and of a higher class in their native land than Mr X
—his father was an architect. As young men, the two of
them acquired a passion for the works of Samuel Butler,
who from 1859 to 1864 had owned and run a sheep farm in
the province of Canterbury, New Zealand. They conceived
the idea of starting a Samuel Butler museum and they wrote
to Festing Jones, Butler's friend and the high priest of his
memory, asking whether he would send them some relics
of Butler. He sent them a few things which became the
nucleus of their museum.

Some time before his call on me Mr X came to England
and became a shop assistant in John Barker in High Street,
Kensington. After a bit Mr Y followed him and they set up
house together in Kensington. They wrote to Festing Jones
and he asked the two young men to dinner. Mr Y decided
that he must have a dress suit for the occasion and Mr X got
into some trouble by borrowing one from John Barker.
After this their financial position was precarious and Mr X
earned a living by going abroad from time to time to Bel-

gium and Holland, where, walking from small town to small
town, he gave readings or recitations of English prose and
poetry. Then another catastrophe befell them. Mr Y wrote
some extremely indecent poems which he wanted to get
printed in order to send them, as Christmas cards, to his
friends. So he went to the policeman who stands at the gates
of the House of Commons and asked him whether he could
recommend him a printer. The policeman gave him an
address of a printer in Whitechapel. Mr Y handed the
printer the manuscript and asked him whether he would
give him an estimate for printing a small number of copies.
This printer, I discovered later, had been fined for printing
the indecent poems of D. H. Lawrence, and, presumably on
the principle of once bitten twice shy, he handed the poems
over to the police. The police prosecuted Mr Y for publishing obscene poems and he was convicted and given three
months.

Mr X wanted to appeal against the conviction and
sentence, but had no money; he had been told by someone
that I might be sympathetic and help him to raise the money
for solicitor's and counsel's fees. I went into the whole thing
carefully and came to the conclusion that it was a monstrous
business. I do not think that Mr Y had ever intended to
publish these poems in the usual sense of publication; to
send them to his friends for Christmas was to be more or less
of a joke; but the magistrate held that his handing the manuscript to the printer was technically and legally 'publication'.
But to give a first offender three months' imprisonment for
this seemed to me gross injustice. I knew Mr Y by sight, for
he was a well-known figure in the streets of London, and as
soon as I realized from Mr X's description who he was, I
saw how prejudice would corrupt the incorruptible British
magistrate or judge before whom he might appear. For Mr
Y dressed himself in a long robe with skirts to the ground

and he wore his hair so long that it hung beneath his shoulders.

I had a talk with Jack Hutchinson, the K.C., about it; he too thought the sentence to be monstrous and was willing to appear in the Appeal Court for a very moderate fee. So I went to a solicitor who lodged an appeal and I began the dreary business of sending round the hat to possible sympathizers. I got donations from seven publishers and more than 20 writers. In the end the solicitor's bill was £91, 7s. 0d. and Mr X spent £12, 4s. 0d. so that I had to raise rather more than £100—it cost me personally over £50. The result was extremely unsatisfactory.

The appeal was heard by Lord Chief Justice Hewart and two other judges. In *Sowing* I wrote the following sentences, to which some people have taken exception, but which I still stand by:

> I have always felt that the occupational disease of judges is cruelty, sadistic self-righteousness, and the higher the judge the more criminal he tends to become. It is one more example of the absolute corruption of absolute power. One rarely sees in the faces of less exalted persons the sullen savagery of so many High Court judges' faces. Their judgments, *obiter dicta*, and sentences too often show that the cruel arrogance of the face only reflects the pitiless malevolence of the soul.

I dare say that in private life Gordon Hewart, 1st Baron Hewart, Lord Chief Justice of England, was a nice man, a good husband and father, a good club man,[1] a pleasant man to play a round of golf with. I watched him 'doing justice'

---

[1] It is worth recording that in *Who's Who* the late Lord Hewart used to list his clubs as follows : Athenaeum, Beefsteak, Garrick, Reform, Savage, United Service; Hadley, Littlestone, Moor Park, South Herts, and Woking Golf.

in the Appeal Court for the better part of a day and he seemed to me—and still seems to me—a typical example of a High Court judge suffering from the occupational disease of sadistic, vindictive self-righteousness. His treatment of the unfortunate Mr Y was disgraceful. One side of his judicial behaviour interested me greatly. In Ceylon for three years I had to do a considerable amount of work as judge and police magistrate and I noticed in myself a curious psychological phenomenon against which one had to be on one's guard if one wanted to be absolutely unprejudiced and just. If one had tried three or four cases one after the other in which one had found the accused guilty, one tended to be overlenient to the next case, particularly if one had had a moment of slight hesitation in finding the last man guilty. And vice versa, if one had found four accused in four cases not guilty one after the other, one had to be very much on one's guard against being unconsciously over-severe to or prejudiced against the accused in the next case.

I was interested to detect the same mental process in the Lord Chief Justice, who, however, took no steps to counteract his bias. In the case which he tried before Mr Y's the accused had been convicted of housebreaking. You only had to look at him to see that he was an old lag. The evidence against him was overwhelming; he had many previous convictions. He appealed upon a tenuous technical point and his counsel made a very clever speech. Hewart set aside the conviction, smacking his judicial lips over the absolute justice of British justice—'the appellant has been extremely lucky in having a counsel to put a difficult case so ably. We are giving the accused, whose record is a bad one, another chance and we hope that he will take it in order to amend his ways. . . .' When I heard this, I felt in my bones that British justice having been so magnanimous to the old burglar would probably take it out on Mr Y. It did. Jack Hutchinson made a good speech and showed to any unbiased person that the sentence was monstrously excessive in relation to the offence. But Hewart made it pretty obvious that he was against Jack and did not like Mr Y. As soon as the case was closed he turned to the judge on his right and to the judge on his left and muttered something to each in turn. The judge on his right was the equivalent on the bench to the old lag, the burglar, whom we had seen in the dock. He had sat on the bench for so long that he administered justice like a machine and therefore mechanically agreed with the Lord Chief Justice. The other on the left was sitting for the first time in the Court of Appeal; I cannot remember his name, but he was a comparatively young man and I knew him to have a reputation for being civilized. He was obviously arguing with Hewart. The three went into a huddle and after a bit the young judge got up and walked round to stand between Hewart and the old judge so that he could put his view more audibly to the old man, as it seemed to me. It was fascinating to watch the, to me, of course, inaudible judicial argument. Hewart was obviously determined and impatient, and at last the young judge with what seemed to me a shrug walked back to his seat. Hewart rejected the appeal with the same self-righteous self-satisfaction with which he had allowed the appeal of the burglar.

Not all the work on the *Nation* was as interesting as this. In fact, after about five years of it, I began to grow rather restive. To be on the editorial side of a journal like the *Nation* is a curious occupation. There is first one's relations with one's immediate equals and superiors. The editorial staff in Great James Street consisted of Hubert Henderson, the editor, Harold Wright, assistant editor, and myself, literary editor. As I have already explained I had made it a condition that within the literary half of the paper I should enjoy practical autonomy. But of course autonomy in that kind of occupation must be relative and limited. I knew Hubert Henderson and liked him. I suppose that in politics in the broadest sense our outlook had a good deal in common. The world is still deeply divided between those who in the depths of their brain, heart, and intestines agree with Pericles and the French revolution and those who consciously or unconsciously accept the political postulates of Xerxes, Sparta, Louis XIV, Charles I, Queen Victoria, and all modern authoritarians. Hubert and I were both on the side of Pericles, but in our interpretation of what liberty, equality, and fraternity meant, or ought to mean, we differed pretty deeply and pretty often. That he voted for the Liberal and I for the Labour Party was in the 1920s not without significance. I thought, and still think, that Liberals after 1914 ought to have realized that Liberalism, like patriotism, is not enough, and that the great problem was to develop an economic liberalism and liberal socialism.

Hubert's articles in the first part of the paper naturally, therefore, often seemed to me extremely able, but conscientiously to hit the bull's-eye on the wrong target. He, on the other hand, regarded a good deal of what I was doing in the other half of the paper with some misprision. Like many liberal intellectuals, he mistakenly prided himself on being culturally an 'ordinary person', a good philistine, with no use for highbrows. This was an amiable delusion, but the result of it was that he had to convince himself that what he called Bloomsbury was impossibly highbrow and that there was far too much Bloomsbury in the literary section of his paper. This worried him more than it did me, for temperamentally, particularly in business and practical affairs, I tend to go my own way and do not worry. Moreover, in my experience there was almost always a pretty deep gulf between the political and literary editors of left-wing weekly papers, like the *Nation* and the *New Statesman*. The cultural scission or schism which, rather to my amusement, I found every now and again opening on a Wednesday morning between me on the one side and Hubert and Harold Wright on the other seemed to me no wider or more catastrophic than that which I had observed between the literary editor of the *New Statesman*, Jack Squire or Desmond MacCarthy, and the editor, Clifford Sharp.

Though this kind of thing did not worry me, it was one example of a good deal of work on the *Nation* which I found more and more boring. In the editorial rooms of weekly newspapers there is or was an unending struggle for space. As soon as the number of pages or columns in the next issue has been settled, the number of pages or columns to be assigned to the editor, the literary editor, and the assistant editor has to be settled. This too often led to a violent struggle for space between politics and literature. Whether the editor should sacrifice a page in which he wanted an article on the Revolution in Bulgaria by Arnold Toynbee to me for an article on John Donne by T. S. Eliot, or vice versa, might well entail a stubborn conflict. This kind of thing is bound up with one side of journalism which, as I have said before, corrodes and erodes the editorial mind. The moment you as editor send back to the printer, say on Wednesday afternoon, the last page proof corrected for the issue of January 1, you have to begin to think of what you are going to put in the issue of January 8, and you have to get that question practically settled by next Monday morning. You are perpetually thinking in terms of articles, notes, reviews, authors, and titles in relation to pages, columns, lines, and words, and the scale of time against which you think of a revolution in Bulgaria or John Donne, of Hitler's Nuremberg Laws or the behaviour of the crowd at the Derby, is five, or at most seven, days. The editorial mind thinks kaleidoscopically in a framework of a few hours or days, but the human mind should, I think, every now and again think steadily *sub specie aeternitatis*. But eternity to the editor of *The Times* is 24 hours and to the editor (and literary editor) of the *Nation* seven days.

Every now and again I have an intense desire for solitude, to shut the door, pull down the blinds, and to be entirely alone for a day or two. When the desire comes upon me, even if Shakespeare or Montaigne knocked at the door, I should pretend to be out. The frame of mind is connected, I think, with a desire occasionally to think of things *sub specie aeternitatis*, the eternal frame of eternity, not the eternity of 24 hours or seven days. To live perpetually in a kaleidoscope of which the kaleidoscopic changes are always more or less the same bores and depresses me. After four years as literary editor of the *Nation* I already began to feel that I had had enough of this kind of journalism and talked to Maynard about giving it up. He wanted me to stay on and eventually I agreed provided that it was arranged that I spent less time in the office, my salary being reduced from £400 to £250. I went to the office only on Tuesdays and Fridays; in fact, I did practically the same work as I had done before—I do not know how I contrived to get through it in less than two full days at the office. But I continued to do this for another three years. In 1929 I told Maynard that I could not stand any more of it and resigned early in 1930.

My resignation from the *Nation* was made possible by our financial situation which was revolutionized in the years 1928 to 1931. From 1924 to 1928 our income only just covered our expenditure and we had to be very careful about both. I propose to give some exact and detailed figures; the private finances of people seem to me always interesting; indeed they have so great an effect upon people's lives that, if one is writing a truthful autobiography, it is essential to reveal them. Here, at any rate are a few figures:

Income in £'s

	LW	Hogarth Press	VW	Other	Gross Income	Tax	Net Income	Expend-iture
1924	569	3	165	310	1,047	126	921	826
1925	565	73	223	404	1,265	114	1,151	846
1926	499	27	713	419	1,658	144	1,514	962
1927	352	27	748	369	1,496	183	1,313	1,193
1928	394	64	1,540	347	2,345	268	2,077	1,117
1929	357	380	2,936	323	3,996	859	3,137	1,120
1930	383	530	1,617	345	2,875	796	2,079	1,158
1931	258	2,373	1,326	411	4,368	1,376	2,992	1,224
1932	270	2,209	2,531	321	5,331	1,278	4,053	1,153
1933	263	1,693	1,916	327	4,199	1,262	2,937	1,187
1934	202	930	2,130	353	3,615	1,086	2,529	1,192
1935	208	741	801	458	2,208	683	1,525	1,253
1936	263	637	721	477	2,098	683	1,415	1,230
1937	271	77	2,466	524	3,184	315	2,869	1,122
1938	365	2,442	2,972	570	6,349	2,462	3,887	1,116
1939	778	350	891	802	2,821	974	1,849	1,069

Some explanation of these figures is necessary. First as regards expenditure. At the end of the 1914 war I invented a system with regard to our finances which we found both useful and amusing and which we kept going until Virginia's death. At the end of each year I worked out in detail an estimate of expenditure for the coming year. This was to provide for only the bare joint expenses of our common life together; it therefore covered rents, rates, upkeep of houses, fuel and lighting, food, servants, garden, upkeep of car (when we got one), doctors and medicine, an allowance to each for clothes. At the end of the year I worked out what the actual expenditure had been and also the total actual combined income, and then the excess of income over expenditure was divided equally between us and became a personal 'hoard', as we called it, which we could spend in any way we liked. For instance, when we decided to have a car, I bought it out of my 'hoard', and if Virginia wanted a new dress which she could not pay for out of her allowance, she paid for it out of her hoard. The amount of our hoards varied enormously as time went on. To take an example, the above figures show that at the end of 1927 we each got £60, but at the end of 1929 we each got £1,008.

The revolutionary increase in our income was due first to the sudden success of Virginia's books and secondarily to the Hogarth Press. In January 1925 Virginia was 42 years old. She had already published three novels (*The Voyage Out, Night and Day,* and *Jacob's Room*) and a book of short stories (*Monday or Tuesday*). In 1924 her income from her books was £37, £21 from her English and £16 from her American publishers; she earned £128 by journalism, so that her total earnings for 1924 were £165. In 1925 she published *Mrs Dalloway* and *The Common Reader* with the Hogarth Press in England and Harcourt, Brace in America. These two books brought her in during the two years 1925 and 1926 £162 in England and £358 in America. In England *Mrs Dalloway* sold 2,236 copies and *The Common Reader* 1,434 copies in the first twelve months. In 1927 *To the Lighthouse* was published and was distinctly more successful than any of her previous books, at any rate in England, where the Hogarth Press sold 3,873 copies in the first year, and she earned from her books in that year £270 in England and £275 in America. That meant that at the age of 47, having written for at least 27 years and having produced five novels, Virginia for the first time succeeded in earning as much as £545 from her books in a year—the most that she had ever made before was £356 in 1926.

The turning-point in Virginia's career as a successful novelist came in 1928 with the publication of *Orlando*. In the first six months the Hogarth Press sold 8,104 copies, over twice as many as *To the Lighthouse* had sold in its first 12 months, and Harcourt, Brace sold 13,031 copies in the first six months. In America Mr Crosby Gaige published a limited edition of 872 copies about a week before the

Harcourt, Brace edition was published. The effect upon Virginia's earnings as a novelist was immediate. In 1928 she earned from her books £1,434 (£556 in England and £878 in America) and in 1929 £2,306 (£761 in England and £1,545 in America). In 1929 *A Room of One's Own* was published and in 1931 *The Waves*. In the first six months *A Room of One's Own* sold 12,443 copies in England and 10,926 in America and *The Waves* sold 10,117 copies in England and 10,380 in America. Virginia's earnings from her books were in 1930 £1,294 (£546 in England and £748 in America), in 1931 £1,266 (£798 in England and £468 in America), and in 1932 £1,795 (£554 in England and £1,241 in America). In 1932 *The Common Reader Second Series* was published and in the first six months sold 3,373 copies in England and 3,271 in America. Then in 1933 came *Flush*. This was a great success. The Hogarth Press sold 18,739 copies in the first six months and Harcourt, Brace 14,081 in America, where it was an alternative selection of The Book-of-the-Month Club.

Virginia finished writing *Flush* in January 1933 and immediately began to work seriously on *The Years*, which she first called *The Pargiters*. It took her four years to write *The Years*, and no major book of hers was published between 1933 and 1937. Her earnings from books during those four years were as follows:

	England	America	Total
1933	£1,193	£1,253	£2,446
1934	301	778	1,079
1935	214	297	511
1936	158	476	634

The Years was published in March 1937 and was much the most successful of all Virginia's books. It was the only one which was a best-seller in America. Harcourt, Brace sold 30,904 copies and in the Hogarth Press we sold 13,005

in the first six months. Virginia's earnings from books for
the years 1937, 1938, and 1939 were as follows:

	England	America	Total
1937	£1,355	£2,071	£3,426
1938	1,697	1,275	2,972
1939	193	254	447

These figures may seem too dull and detailed to many
people, but autobiographically and biographically they are
important. The facts behind them had economically a
considerable effect upon our lives. After 1928 we were
always very well off. In the next ten years our income was
anything from twice to six times what it had been in 1924.
Neither of us was extravagant or had any desire for con-
spicuous extravagance; we did not alter fundamentally our
way of life, because on £1,000 a year we already lived the
kind of life we wished to live, and we were not going to
alter the chosen pattern of our life because we made £6,000
in the year instead of £1,000. But life is easier on £3,000
a year than it is on £1,000. Within the material framework
which we had chosen for our existence we got more of
the things which we liked to possess—books, pictures,
a garden, a car—and we did more of the things we wanted
to do, for instance travel, and less in the occupations which
we did not want to do, for instance journalism.

But the statistics of Virginia's earnings as a writer of books
have from another point of view still greater interest and
importance. They throw a curious light on the economics of
a literary profession and on the economic effect of popular
taste on a serious writer. *Orlando*, *Flush*, and *The Years* were
immeasurably more successful than any of Virginia's other
novels. *The Years*, much the most successful of them all, was,
in my opinion, the worst book she ever wrote—at any rate, it
cannot compare, as a work of art or a work of genius, with

The Waves, *To the Lighthouse*, or *Between the Acts*. *Orlando* is a highly original and amusing book and has some beautiful things in it, but it is a *jeu d'esprit*, and so is *Flush*, a work of even lighter weight; these two books again cannot seriously be compared with her major novels. The corollary of all this is strange. Up to 1928, when Virginia was 46, she had published five novels; she had in the narrow circle of people who value great works of literature a high reputation as one of the most original contemporary novelists. Thus her books were always reviewed with the greatest seriousness in all papers which treat contemporary literature seriously. But no one would have called her a popular or even a successful novelist, and she could not possibly have lived upon the earnings from her books. In 1932 Mrs Leavis, rather a hostile critic, wrote:

> The novels are in fact highbrow art. The reader who is not alive to the fact that *To the Lighthouse* is a beautifully constructed work of art will make nothing of the book. . . . *To the Lighthouse* is not a popular novel (though it has already taken its place as an important one), and it is necessary to enquire why the conditions of the age have made it inaccessible to a public whose ancestors have been competent readers of Sterne and Nashe.[1]

Mrs Leavis exaggerates. It is not true, as the subsequent history of *To the Lighthouse* shows, that the 'common reader' who does not bother his head about 'beautiful construction' or indeed works of art, can make nothing of the book. There is no reason to think that *Tristram Shandy* was more 'accessible' to the eighteenth-century common reader than *To the Lighthouse* is to the twentieth-century common reader. Mrs Leavis in another passage even more strangely asserts that *To the Lighthouse* is more highbrow art and less accessible to the

[1] *Fiction and the Reading Public*, by Q. D. Leavis, p. 223.

ordinary person than Henry James's *Awkward Age* and *The Ambassadors*. But it is, of course, true, as I have shown above by the statistics that up to 1928 Virginia, although widely recognized as an important novelist, was read by a small public. The fate of her books after 1928, however, points to a conclusion quite different from, and more interesting than, Mrs Leavis's. Take, for instance, the sales of *To the Lighthouse* after 1928 and up to the present date. By 1964 the book had sold 113,829 copies in Britain and 139,644 copies in America. It is selling more today than it has ever sold since its publication in 1927. For instance, in 1964 it sold 10,142 copies in Britain and 13,060 in America, and in 1965 22,340 in Britain and 21,309 in America. A book which sells 43,649 copies a year 39 years after publication cannot be said to be unpopular or un-understandable by ordinary people.

A graph of the sales of Virginia's books and of her reputation since 1920 suggests that, so far as original writing is concerned, the law with regard to literature is the exact opposite of Gresham's law by which bad money drives out good. Nearly all artists, from Beethoven downward, who have had something highly original to say and have been forced to find a new form in which to say it, have had to pass through a period in which the ordinary person has found him unintelligible or 'inaccessible', but eventually, in some cases suddenly, in others gradually, he becomes intelligible and is everywhere accepted as a good or a great artist. In Virginia's case she had to write a bad book and two not very serious books before her best serious novels were widely understood and appreciated. And in her case the good drove out the bad. In the years 1963 and 1964, when *To the Lighthouse* sold annually 23,000, the sales of *The Years* and *Orlando* were negligible. In America they were out of print, and in Britain *Orlando* sold 641 in 1963 and 509 in 1964, *The Years* 213 in 1963 and 470 in 1964. But *The Waves*, the most difficult and the best of all her books, sold 906 in 1963 and 1,336 in 1964, and *Mrs Dalloway*, another difficult and the most 'highbrow' of her books, sold 8,242 copies in 1963 (2,306 in Britain and 5,936 in America) and 10,791 in 1964 (2,098 in Britain and 8,693 in America).

In *A Writer's Diary* I published extracts from Virginia's diary which show her engrossed in the day-to-day work of writing these books. She uses these pages as Beethoven used his Notebooks to jot down an idea or partially work out a theme to be used months or years later in a novel or a symphony. While writing a book, in the diary she communes with herself about it and its meaning and object, its scenes and characters. She reveals, more nakedly perhaps than any other writer has done, the exquisite pleasure and pains, the splendours and miseries, of artistic creation, the relation of the creator both to his creation and his creatures and also to his critics and his public. Her hypersensitivity—the fact that criticism tortured her mind like the dentist's drill on an exposed nerve—has seemed to many of her posthumous critics extraordinary and highly discreditable. She herself agreed with her critics that it was highly discreditable.[1] No

---

[1] On May 17, 1932, Virginia wrote in her diary: 'What is the right attitude towards criticism? What ought I to feel and say when Miss B. devotes an article in *Scrutiny* to attacking me? She is young, Cambridge, ardent. And she says I'm a very bad writer. Now I think the thing to do is to note the pith of what is said—that I don't think—then to use the little kick of energy which opposition supplies to be more vigorously oneself. It is perhaps true that my reputation will now decline. I shall be laughed at, pointed at. What should be my attitude—clearly Arnold Bennett and Wells took the criticism of their youngers in the wrong way. The right way is not to resent; not to be longsuffering and Christian and submissive either. Of course, with my odd mixture of rashness and modesty (to analyse roughly) I very soon recover from praise and blame. But I want to find out an attitude. The most important thing is not to think very much about oneself. To investigate candidly the charge; but not fussily, not very

doubt it was partly due to the fact, which I have noted before, that 'there was too much ego in her cosmos', and an excess of egoism is discreditable. It was also partly due to her attitude to her work, her art, her books. The vast majority of people work for about eight hours a day, and during those eight hours apply less than 50 per cent. of their attention or concentration to the work. Out of the 16 hours of her waking day I should reckon that Virginia normally 'worked' 15 hours and I should guess that she dreamed about it most of the time when she was asleep. Her work was her writing, and when she was actually writing, her concentration was 100, not 50, per cent. But unlike most people, she was almost always at her work even when she was not working. Practically every afternoon, when at Rodmell, she would walk for an hour, two hours, or even more. All the time on the downs, across the water-meadows, or along the river bank, in the front or at the back of her mind was the book or article she was writing or the embryo of a book or story to be written. It was not that she did not see or feel her surroundings, the kaleidoscope of fields, downs, river, birds, a fox or a hare—she saw and felt them with intensity, as her conversation and the extraordinarily visual imagery of her writing show; but at the same time, at the back or just below the surface of her mind there seemed to be a simmering of thoughts, feelings, images connected with her writing and every now and again this simmering would rise to the surface or boil over in the form of a conscious consideration of

---

anxiously. On no account to retaliate by going to the other extreme—thinking too much. And now that thorn is out—perhaps too easily.' The word 'thorn' here has a kind of special meaning. We used to say that Virginia was continually picking up mental thorns—worries which she could not get rid of—particularly from criticism. She would come to me and say: 'I've got a thorn', and we would discuss the thing until we had got the thorn out.

a problem in her book or the making of a phrase or the outline of a scene to be written next morning.

Moreover, though she enjoyed intensely for their own sake the sights and sounds of her walk, these too were, I think, almost always registered as to some extent the raw material of her art. The same is true of all the activities of her life. For instance, as I have said before, no one could possibly have enjoyed society or parties more than she did; on the surface she was carried away by them and so more often than not was a great 'social success'. And yet I do not think that that second sight, that second layer in her mind, was ever entirely quiescent; there the scene, the dinner-party, the conversation, her own feelings were continually registered and remembered as the raw material of her art. This is shown by the fact that she so often described parties in considerable detail in her diary.[1]

[1] The following is an example written on May 26, 1932: 'Last night at Adrian's party. Zuckerman on apes. Dora Chapman sitting on the floor. I afraid of Eddy (Eddy Sackville-West) coming in—I wrote him a sharp but well earned letter. Adrian so curiously reminiscent—will talk of his school, of Greece, of his past as if nothing had happened in between: a queer psychological fact in him—this dwelling on the past, when there's his present and his future all round him: D. C. to wit and Karin coming in late, predacious, struggling, never amenable or comforting as, poor woman, no doubt she knows: deaf, twisted, gnarled, short, stockish, baffled, still she comes. Dick Strachey. All these old elements of a party not mingling. L. and I talk with some effort. Duncan wanders off. Nessa gone to Tarzan. We meet James and Alix on the door. Come and dine, says James with the desire strong in him I think to keep hold of Lytton. Monkeys can discriminate between light and dark: dogs can't. Tarzan is made largely of human apes. . . . Talk of Greece. Talk of Spain. Dick was taken for a ghost. A feeling of distance and remoteness. Adrian sepulchral, polite, emaciated, elongated, scientific, called Adrian by Solly; then in come rapid small women, Hughes and I think his wife. We evaporate at 11.20: courteously thanked for coming by Adrian. Question what pleasure these parties give. Some, presumably, or these singular figures wouldn't coagulate.'

This unremitting intensity with which she worked upon whatever she was writing, when combined with her sensitiveness to all sensations and impressions, was perhaps to some extent a cause of her vulnerability, the intensity of her feeling about criticism of it. Even when she had reached the dangerous pinnacle of success, established as an important writer, she never showed the slightest sign of that fatal occupational disease of the successful writer, the feeling of being a very important person. On the contrary, the more successful she became, the more vulnerable she seemed to become, with a kind of humility and uncertainty which were the exact opposite of the assurance and importance which one felt in the great men of their day, like Wells, Bennett, Galsworthy, and Shaw,[1] and even in many of the smaller fry.

The implacable intensity of concentration upon her writing and her almost pathological fear of the exposure of publication combined to produce the exhaustion and despair which assailed her in the interval between finishing a book and publishing it. All the books, from *Jacob's Room* to *The Years*, induced one of these dangerous crises. I have already referred to them and I do not propose to record in any detail the effect of each, great though it was, upon Virginia herself and upon our day-to-day life. The exact nature of them can perhaps best be seen in what she wrote on December 23, 1932, when she had just finished *Flush*, a book which she wrote very easily and never took too seriously:

I must write off my dejected rambling misery—having just read over the 30,000 words of *Flush* and come to the conclusion that they won't do. Oh what a waste—what a bore! Four months of work and heaven knows how much reading—not of an exalted kind either—and I can't see how to make anything of it. It's not the right subject for

---

[1] But characteristically not in Hardy.

that length: it's too slight and too serious. Much good in it but would have to be much better. So here I am two days before Christmas pitched into one of my grey welters.

Her major works, by increasing the strain, only increased that 'dejected rambling misery' and the 'grey welter'. The moment she sent back the corrected proofs of *The Waves* to the printer she had to go to bed with a dangerous headache. When the book was published and before she had any criticism of it, whether from Hugh Walpole, John Lehmann, or anyone else, she wrote in her diary: 'I have come up here, trembling under the sense of complete failure—I mean *The Waves*—I mean Hugh Walpole[1] does not like it—I mean John L. is about to write to say he thinks it bad—I mean L. accuses me of sensibility verging on insanity'. And months later she said that she still felt her brain numb from the strain of writing *The Waves*.

How near these strains from writing and publishing brought her at any moment to breakdown and suicide is shown frequently in her diary. For instance, early in July 1933 the worry over revising *Flush* and her excitement over beginning to write *The Years* brought on a headache, and she recorded on July 10: 'And then I was in "one of my states" —how violent, how acute—and walked in Regents Park in black misery and had to summon my cohorts in the old way to see me through, which they have done more or less. A note made to testify to my own ups and downs: many of which go unrecorded though they are less violent I think than they used to be. But how familiar it was—stamping

[1] She noted herself later that, though Hugh said that it was 'unreal' and that it beat him, John 'loved it, truly loved it, and was deeply impressed and amazed'. And she adds about herself: 'My brain is flushed and flooded. . . . Lord what a weathercock—not a wave of emotion is in me.'

along the road, with gloom and pain constricting my heart: and the desire for death, in the old way, all for two I dare say careless words.'

Although her day-to-day mental health in general became stronger and more stable through the 1920s and 1930s, the crises of exhaustion and black despair when she had finished a book seemed each time to become deeper and more dangerous. We had a terrifying time with *The Years* in 1936; she was much nearer a complete breakdown than she had ever been since 1913. There are two gaps in her 1936 diary, one of two months between April 9 and June 11 and another of four months between June 23 and October 30. They were filled with an unending nightmare. For the first three months of the year Virginia was revising the book and, as she revised it, we sent it to the printer to be put into galley proofs. We did this—getting galley instead of page proofs—because Virginia was in despair about the book and wanted galleys so that she would be free to make any alterations she wished in proof. But at the beginning of May she was in such a state that I insisted that she should break off and take a complete holiday for a fortnight. We drove down into the west country by slow stages, stopping in Weymouth, Lyme Regis, and Beckey Falls on Dartmoor, until we reached Budock Vean in that strange primordial somnolent Cornish peninsula between Falmouth and Helford Passage, where the names of the villages soothe one by their strangeness—Gweek and Constantine and Mawnan Smith. As a child Virginia had spent summer after summer in Leslie Stephen's house at St Ives in Cornwall—the scene of *To the Lighthouse* is St Ives and the lighthouse in the book is the Godrevy light which she saw night by night shine across the bay into the windows of Talland House. No casements are so magic, no faery lands so forlorn as those which all our lives we treasure in our memory of the summer holidays of our childhood.

Cornwall never failed to fill Virginia with this delicious feeling of nostalgia and romance.

I thought that for Virginia's jangled nerves I might find in Cornwall the balm which the unfortunate Jeremiah thought—mistakenly—he might find in Gilead to salve the 'hurt of the daughter of my people'. That was why I drove west and stayed in Budock Vean, and revisited Coverack and the Lizard and Penzance, and went on to stay with Will and Ka Arnold-Forster in that strange house, Eagle's Nest, perched high up on the rock at Zennor a few miles from St Ives. As the final cure, we wandered round St Ives and crept into the garden of Talland House and in the dusk Virginia peered through the ground-floor windows to see the ghosts of her childhood. I do not know whether, like Heine, she saw the Doppelgänger and heard the mournful echo of Schubert's song: 'Heart, do you remember that empty house? Do you remember who used to live there? Ah, someone comes! Wringing her hands! Terrible! It is myself. I can see my own face. Hi, Ghost! What does it mean? What are you doing, mocking what I went through here all those years ago'.

I drove back by easy stages to Rodmell and then to London. Virginia seemed to be a good deal better, and, as a further precaution, she took another twelve days' complete holiday at Rodmell. She started to work on the proofs again on June 12, but almost at once it became clear that she had not really recovered. After nine or ten days we decided that she must break off altogether and take a complete rest. In fact, on July 9 we went down to Rodmell and stayed there for three and a half months. Virginia did not write at all, did not look at her proofs, and hardly ever moved out of Rodmell. Once a week I drove up to London for the day and it was a pretty strenuous day. I used to leave Rodmell at 8 and get to Tavistock Square about 10, where I dealt with the Hogarth

Press business. In the afternoon I went to the House of Commons for a meeting of the Labour Party Advisory Committee. That meant that quite often I did not leave London until nearly 7 so that I did not get back to Rodmell until close on 9. Virginia did not accompany me on these weekly expeditions. She spent her time reading, drowsing, walking. Towards the end of October she seemed very much better and we decided that I should read the proofs of *The Years* and that she would accept my verdict of its merits and defects and whether it should or should not be published. It was for me a difficult and dangerous task. I knew that unless I could give a completely favourable verdict she would be in despair and would have a very serious breakdown. On the other hand, I had always read her books immediately after she had written the last word and always given an absolutely honest opinion. The verdict on *The Years* which I now gave her was not absolutely and completely what I thought about it. As I read it I was greatly relieved. It was obviously not in any way as bad as she thought it to be; it was in many ways a remarkable book and many authors and most publishers would have been glad to publish it as it stood. I thought it a good deal too long, particularly in the middle, and not really as good as *The Waves*, *To the Lighthouse*, and *Mrs Dalloway*.

To Virginia I praised the book more than I should have done if she had been well, but I told her exactly what I thought about its length. This gave her enormous relief and, for the moment, exhilaration, and she began to revise the proofs in order to send them back to the printer. She worked at them on and off from November 10 until the end of the year, sometimes fairly happy about the book and sometimes in despair. 'I wonder', she wrote in her diary, 'whether anyone has ever suffered so much from a book as I have suffered from *The Years*'; I doubt whether anyone ever has.

But—how often in my life I have gratefully murmured Swinburne's lines—'even the weariest river winds somewhere safe to sea'. She revised the book in the most ruthless and drastic way. I have compared the galley proofs with the published version and the work which she did on the galleys is astonishing. She cut out bodily two enormous chunks, and there is hardly a single page on which there are not considerable rewritings or verbal alterations. At last on December 31 the proofs were returned to the printer. The book was published in March of the following year, and, as I have said, at once proved to be the greatest success of all the novels which she had written.

Virginia was a slow writer. It took her four years to write *The Years* and there was an interval of six years between *The Waves* and *The Years*. Yet she was comparatively a prolific writer. She wrote nine full-length novels, two biographies, and there are seven volumes of literary criticism; in addition to this there must be at least 500,000 words of her unpublished diaries. As a novelist her output was greater than that of Fanny Burney, Jane Austen, the Brontës, George Eliot, Thackeray, or in modern times Joyce and E. M. Forster. This is a remarkable fact when one thinks of the psychological handicaps and difficulties which I have described in the previous pages. It was to a great extent due to her professional, dedicated, industriousness. Neither of us ever took a day's holiday unless we were too ill to work or unless we went away on a regular and, as it were, authorized holiday. We should have felt it to be not merely wrong but unpleasant not to work every morning for seven days a week and for about eleven months a year. Every morning, therefore, at about 9.30 after breakfast each of us, as if moved by a law of unquestioned nature, went off and 'worked' until lunch at 1. It is surprising how much one can produce in a year, whether of buns or books or pots or pictures, if one works hard and professionally for three and a half hours every day for 330 days. That was why, despite her disabilities, Virginia was able to produce so much.

Thus, although she was in such a desperate state about *The Years* all through the last six months of 1936, she already had simmering in her mind *Three Guineas* and the biography of Roger Fry. Indeed on January 28, 1937, exactly one month after sending back the proofs of *The Years* to the printer, she began to write *Three Guineas*; she finished the first draft on October 12, 1937, and began to write *Roger Fry* on April 1, 1938. But already in August 1937 a new novel, which was to become *Between the Acts*, was simmering in her mind and she began to write it, under its first title *Poyntz Hall*, in the first half of 1938. In 1938 we published *Three Guineas* and in 1940 *Roger Fry*. On February 26, 1941, she finished *Between the Acts* and, as had happened four years before, she fell into the depths of despair. On March 28 she drowned herself in the Ouse.

I must return to the subject of our income. I said that the sudden, large jump upwards in our income which began in 1928 was primarily due to the sudden success of Virginia's books. But I also said that it was due secondarily to the Hogarth Press and I will deal now with the development of this curious publishing business. When we moved from Richmond to Tavistock Square in March 1924, the Press, though it had published 13 books in the previous 12 months, was still a very amateurish affair. It had one employee, Marjorie Joad. 1924 was again a year of considerable expansion and we had three publications which in particular were to have a great influence upon our future as publishers.

In 1924 Vita asked us whether we would like to publish a longish short story which she had written, *Seducers in Ecuador*. At that time she had already published some poems and two or three novels with Collins and Heinemann.

Seducers was a curious little story which no ordinary publisher would have looked at. We made a very pretty little book out of it and published it at 4s. 6d. just before Christmas. When we sold out the edition of 1,500 copies we did not reprint. At that time Harold was still in the diplomatic service and in 1925 he was appointed to the British Legation in Teheran. Vita went with him and she wrote a good travel book about Persia, *Passenger to Teheran*, which we published in 1926. She followed this up with *Twelve Days* in which she described an adventurous journey across the Bakhtiari mountains to the Persian oil-fields, and we published this book in 1928. Next year she brought us the manuscript of *The Edwardians*. This was a novel about Knole, the Sackvilles, and Edwardian Society with the most aristocratic capital S, written from the inside of not only Knole, but also Vita. Inside Vita was an honest, simple, sentimental, romantic, naïve, and competent writer. When she let all this go off altogether in a novel about high life, she produced in *The Edwardians* a kind of period piece and a real best-seller. Both Virginia and Vita had been warned by friends and friendly publishers that it was madness to have their books published by such an amateurish, ramshackle concern as the Hogarth Press, which had not the machinery to deal with a best-seller or even a seller. I have always been doubtful about this 'machinery' of publishing and was pleased to find that the machinery of the Press stood up to the strain of a best-seller. We sold nearly 30,000 copies of *The Edwardians* in the first six months, and by the end of a year the Press had made a profit of nearly £2,000 on it. It has gone on selling for years.

Novels by serious writers of genius often eventually become best-sellers, but most contemporary best-sellers are written by second-class writers whose psychological brew contains a touch of naïvety, a touch of sentimentality, the story-telling gift, and a mysterious sympathy with the day-dreams of ordinary people. Vita was very nearly a best-seller of this kind. She only just missed being one because she did not have quite enough of the third and fourth element in the best-selling brew. We published *The Edwardians* in 1930 and *All Passion Spent*, which she wrote in less than a year, in 1931. This was, I think, the best novel which she ever wrote, though there was rather more than a touch of sentimentality in it. It did very well, though not as well as *The Edwardians*, selling about 15,000 copies in the first year —it still sells 35 years after it was first published—and showing a profit of about £1,200. After this book the springs of Vita's invention and imagination which she required for novel writing began to run dry. She produced a fascinating and very amusing book, *Pepita*, a biography of her terrible mother and extraordinary grandmother, which we published with great success in 1937. But I had grave doubts about her two novels, *Family History* and *The Dark Island*, and then she brought us the manuscript of a novel which we felt we could not publish.

The relation of author and publisher is never an easy one. The publisher is, at the best of times, an ambivalent and often not very competent business man, wobbling between profits and art for art's sake; the writer has something of the same kind of wobble and is often convinced that the reason why his book is not a best-seller is because his publisher is an incompetent, profit-making shark. Vita was an ideal author from the publisher's point of view; she never complained when things went wrong and was extraordinarily appreciative of the publisher if they went right. This made it all the more unpleasant to have to tell her that we thought her novel not good enough for us to publish. We knew, too, that we should lose her as an author, because there were many reputable publishers who would publish this novel in order to get her 'on their list'. It was characteristic of her that she was not in the least bit hurt or resentful and the whole thing made no difference to her relationship with us.

By 1924, seven years after it started, the Hogarth Press, still with practically no employees, no capital invested, and no overheads, had on its general list already two potentially best-sellers, Vita and Virginia. But we were not merely a one- or two-horse shay. With courageous wisdom or reckless folly, we took on a considerable number of new authors and new books. There were 34 books announced in our 1925 lists and we had published all of them by 1926. In the next five years we published novels by William Plomer, Edwin Muir, F. L. Lucas, C. H. B. Kitchin, Alice Ritchie, F. M. Mayor, Svevo, and Rilke. The first six of these were first novels. Two of the six, which did quite well when published, are now forgotten, but are, I still think, remarkable, *The Rector's Daughter* by F. M. Mayor and *The Peacemakers* by Alice Ritchie. A little later, in the early 1930s we published Christopher Isherwood's *Mr Norris Changes Trains*, new novels by Ivan Bunin, and Laurens van der Post's first novel *In a Province*. In general publishing we also branched out into art with a classic, *Cézanne* by Roger Fry, in 1926, and very strongly into politics, economics, history, and sociology. This last category was directly connected with my own activities. In the 1920s I wrote *After the Deluge*, Vol. I, and *Imperialism and Civilization*. But I also became more and more occupied with practical politics in the Labour Party and Fabian Society and this is reflected in the large number of political books which we published, among which some of the most important were *The End of Laissez-Faire* (1926) and *The Economic Consequences of Mr Churchill* (1925) by Maynard Keynes, the remarkable books on imperialism in Africa by Norman Leys, and Lord Olivier's *White Capital and Coloured Labour*.

I said above that there were three publications in 1924 which had a considerable influence on the development of the Press. The first was Vita's book. The second was a series, the Hogarth Essays, which we started in 1924 with three volumes, *Mr Bennett and Mrs Brown* by Virginia, *The Artist and Psycho-Analysis* by Roger Fry, and *Henry James at Work* by Theodora Bosanquet. This series consisted of pamphlets, a form of publication which nearly all publishers fought (still fight) shy of because they always involve a good deal of work and a loss of money. I was eager to have a series of pamphlets in which one could have essays on contemporary political and social problems as well as on art and criticism. These essays, published at between 1s. 6d. and 3s. 6d., were surprisingly successful. In the first series in 1924, 1925, and 1926, we published 19 essays, bound in stiff paper with a cover design by Vanessa, and among the authors were T. S. Eliot, Robert Graves, Edith Sitwell, J. M. Keynes, E. M. Forster, J. A. Hobson, Vernon Lee, Bonamy Dobrée, and Herbert Read. None of them sold a large number of copies, but every one of them, when they went out of print, had made a profit. This encouraged me to start in 1930 another series, Day to Day Pamphlets, devoted entirely to politics, bound in paper, and published at 1s. or 1s. 6d. In the nine years between 1930 and 1939 we published 40 pamphlets, and among the authors were Harold Laski, H. N. Brailsford, W. H. Auden, H. G. Wells, Sir Arthur Salter, C. Day Lewis, A. L. Rowse, and Mussolini. This series also paid its way. It was significant of the political climate in the 1930s that the two best-sellers were Mussolini's *The Political and Social Doctrine of Fascism* and Maurice Dobb's *Russia Today and Tomorrow*, an excellent, though rather rosy, view of Soviet Russia and communism by a Cambridge don.

The pamphlet is not a commodity which it is easy to sell in Britain. Every now and again if you publish one by the

right person at the right moment on some controversial
subject, it will be a best-seller. We printed 7,000 copies of
The Economic Consequences of Mr Churchill by Maynard and
they sold out at once, but in the general run when we were
publishing regularly four or five every year, we did well if
we sold over 2,000 copies. Societies like the Fabians which
regularly published pamphlets found the same difficulties in
selling them. The principal obstruction is the trade. The
pamphlet is an awkward and troublesome kind of book to sell,
and most bookshops will not look at them. The railway book-
stall, which is the right place to market them, also dislikes
them, not without reason, as they are not nearly as lucrative
as newspapers or that vast range of lurid or alluring publications on whose covers are portrayed murder and rape or
the superfeminized female form in every stage of dress or
nudity. The result is that the British have never acquired the
habit of reading pamphlets. This is a great pity. The pamphlet is potentially an extraordinarily good literary form
from both the artistic and the social or political point of
view. Those which we published by T. S. Eliot, Roger Fry,
Virginia, Maynard Keynes, J. A. Hobson, were remarkable
and would never have been written if we had not had this
series. All the others were, I think, well worth publishing.
Our experience showed us that there is a potential market for
the pamphlet. The potentiality never becomes actuality because all avenues of sale between publisher and purchaser
are closed or obstructed. And so the habit of writing and
reading pamphlets cannot establish itself.

However, these pamphlet series were sufficiently successful to pay their way. They showed us how valuable from
the business point of view a series is to a publisher. If one
gets a series started successfully with good books, it makes
it possible subsequently to publish in the series successfully
other books which, if published on their own, however good

they might be, would almost certainly have made a substantial loss. In the next few years we started four other series: Hogarth Lectures on Literature, Hogarth Sixpenny Pamphlets, Hogarth Living Poets, and Hogarth Letters. The Living Poets and the Lectures were highly successful. I was over-optimistic about the other two. They contained some interesting and amusing essays, for instance, *A Letter to Madan Blanchard* by E. M. Forster, *A Letter to a Sister* by Rosamond Lehmann, *A Letter to a Grandfather* by Rebecca West. But it was impossible to sell enough of them at 6d. and 1s. to make both ends meet.

The third publication of the Press in 1924 which was to have a considerable effect on its future was a strange one. The publication was announced in our autumn list as follows:

COLLECTED PAPERS. By Sigmund Freud, M.D.

Vol. I.   Early Papers and the History of the Psycho-Analytical Movement.

Vol. II.   Clinical Papers and Papers on Instinct and the Unconscious.

The Hogarth Press has taken over the publications of the International Psycho-Analytical Library, and will in future continue the series for the International Psycho-Analytical Press. It has obtained the right to publish a complete authorized English translation of Professor Freud's collected papers. These papers are of the highest importance for the study of Psycho-Analysis; they have been translated into English by experts under the supervision of Dr Ernest Jones. The Collected Papers will be published in four volumes. Vol. III, containing 'Five Case Histories', and Vol. IV, containing 'Metapsychology. Dreams', will be published in the course of next year.

The price of Vols. I, II, and IV will be 21s. each, of Vol. III 30s., and of the complete set four guineas.

That was the beginning of our connection with Freud and the Institute of Psycho-Analysis which has lasted until the present day. It came about in the following way. In the decade before 1924 in the so-called Bloomsbury circle there was great interest in Freud and psycho-analysis, and the interest was extremely serious. Adrian Stephen, Virginia's brother, who worked with Sir Paul Vinogradoff on mediaeval law, suddenly threw the Middle Ages and law out of the window, and, with his wife Karin, became a qualified doctor and professional psycho-analyst. James Strachey, Lytton's youngest brother, and his wife also became professional psycho-analysts. James went to Vienna and was analysed by Freud, and he played an active part in the Institute of Psycho-Analysis, which, largely through Ernest Jones, had been founded in London and was in intimate relations with Freud and the Mecca of psycho-analysis in Vienna, being itself a branch of the International Association of Psycho-Analysis.

Some time early in 1924 James asked me whether I thought the Hogarth Press could publish for the London Institute. The Institute, he said, had begun the publication of the International Psycho-Analytical Library in 1921 and had already published six volumes, which included two of Freud's works, *Beyond the Pleasure Principle* and *Group Psychology and the Analysis of the Ego*. They had also signed an agreement with Freud under which they would publish his *Collected Papers* in four volumes. The Institute had hitherto been their own publisher, printing and binding their books in Vienna and having them 'distributed' by a large London publishing firm. They did not find this system satisfactory and they wished to hand over to a publisher the entire business of publishing the International Psycho-Analytical Library in which they hoped regularly to publish a considerable number of important books by Freud and other analysts.

The idea seemed to me very attractive and I drew up an

agreement which the Institute accepted. It was agreed that we should take over the books which they had already published, and publish all future books in the Library. For a fledgling inexperienced publisher this was a bold undertaking. The four volumes of Freud's *Collected Papers* were a formidable work, for each of them ran to over 300 pages and it meant putting a good deal of capital into them. In fact, one of the most distinguished of the large London publishers, who heard what I was about to undertake, gave me a friendly warning that I should be risking too much. The *Collected Papers* was from the start one of the most successful of our publications. I circularized a large number of universities, libraries, and individuals in the United States and had almost at once a good sale in America as well as in Britain. The Institute had bought outright from Freud the rights for £50 for each volume and we bought the rights in the four volumes from the Institute for £200, but as soon as the books began to make a profit, we began to pay a royalty to Freud. The sale of these four fat volumes (we added a fifth later) has gone on steadily for over 40 years. The fact that it was started successfully by a publisher with no staff and no 'machine' throws a curious light upon the business of publishing. The longer my experience of the business, the more convinced I have become that the ideas of most authors and indeed many publishers with regard to the efficacy or necessity of what is called the publisher's 'machinery' for selling books is largely delusion, though this applies only to 'serious' books, not to that branch of large-scale industry and mass production, the bestseller racket, in which books have to be sold by the same methods as beers. You may be able to sell a million copies of Mr X's *Women and Wine* or Miss Y's *Wine and Women* by the pressure cooker of large-scale advertising and the mysterious machinery of the colossal publisher, but it is doubtful whether you will sell ten copies of Freud's *Collected Papers*,

T. S. Eliot's *The Waste Land*, or Virginia Woolf's *The Waves* by these methods.

Publishing the Psycho-Analytical Library for the Institute was always a very pleasant and very interesting experience. In the next 40 years we published nearly 70 volumes in it. In the process I learnt a good many curious things about the art of publishing. For instance, we had in the Library a book by Professor Flügel called *The Psycho-Analytic Study of the Family*. I do not believe that any publisher who saw this book in manuscript or in print or in our list in 1924 would have thought that it had the slightest chance of being a best-seller, and I feel sure that very few of my readers in 1967 have ever heard of it. Yet this book has been a steady seller for over 40 years, selling hundreds of copies yearly. It has practically never been advertised and no advertising would have materially influenced its sale. Its aggregate sale must be considerably greater than that of nine out of ten of the much advertised best-sellers that it has long outlived. It is an original book, an almost unknown classic in its own peculiar field, a publisher's dream. It sold steadily in Britain year after year, and year after year twice a year there came a large order from an American bookseller, because it was a 'set book' in an American college.

The greatest pleasure that I got from publishing the Psycho-Analytical Library was the relationship which it established between us and Freud. Between 1924 when we took over the Library and his death in 1939 we published an English translation of every book which he wrote, and after his death we published his complete works, 24 volumes, in the Standard Edition. He was not only a genius, but also, unlike many geniuses, an extraordinarily nice man. The business connected with his books, when we first began to publish them, was managed by his son Martin, and later after his death by his son Ernst and daughter Anna. They all seemed to have inherited the extraordinarily civilized temperament of their father which made every kind of relationship with him so pleasant.

The publication of the Standard Edition of the Complete Psychological Works of Freud, which we began in 1953, was one of the most difficult and delicate business operations which I have ever put through. The Internationaler Psychoanalytischer Verlag had already in the 1920s published the *Gesammelte Schriften* of Freud in Germany, Austria, and Switzerland, and after the war in 1942 the complete works in German (*Gesammelte Werke*) were published in London. I was most anxious to publish a complete edition in English and in the 1940s discussed the possibility several times with Ernst Freud. There seemed to be insuperable difficulties. Financially an English edition was not feasible unless it could be sold in the United States as well as in the British Commonwealth, but Freud's American copyrights were in such a tangled and chaotic condition that the moment we began definitely to face the problem there seemed to be no way of acquiring all the rights or even of being quite sure of who controlled them. Freud had been incredibly generous and casual with his copyrights. The outright sale of his English rights in the four volumes of *Collected Papers* for £200 was a good example of his generosity; his casualness is shown by the fact that he once simultaneously sold the American rights in one of his books both to the Hogarth Press and to another publisher. In America he had given the copyright in many of his books to his translator and friend, Dr Brill. This—quite apart from the difficulty of discovering who owned the American rights in several of the other books—presented us with an extremely delicate problem. We had been lucky enough to be able to arrange that, if we did ever succeed in publishing the complete works, the edition would be edited and translated by James Strachey, assisted by his wife, Alix. James has, in fact, accomplished this colossal task in 13 years, and the 24 volumes are a monument to his extraordinary combination of psycho-analytical knowledge, brilliance, and accuracy as a writer and translator, and indomitable severity both to himself and to his publisher. I doubt whether there is any edition of technical scientific works, comparable to this one in size, which can compare with it in the high standard of translation and editing.

What made the task of our approach to Dr Brill, and after his death to his heirs and executors, doubly delicate was that we had to obtain their consent first to our publishing in America translations of the works in which they held the American rights, and second to our using not Dr Brill's translations but James's. For a long time every attempt to find a way through the maze of copyrights and overcome the other difficulties failed, but eventually, largely owing to Ernst Freud's tact and perseverance, an agreement with the Brill executors made publication in America possible.

I only once met Freud in person. The Nazis invaded Austria on March 11, 1938, and it took three months to get Freud out of their clutches. He arrived in London in the first week in June and three months later moved into a house in Maresfield Gardens which was to be his permanent home. When he and his family had had time to settle down there, I made discreet enquiries to see whether he would like Virginia and me to come and see him. The answer was yes, and in the afternoon of Saturday, January 28, 1939, we went and had tea with him. I feel no call to praise the famous men whom I have known. Nearly all famous men are disappointing or bores, or both. Freud was neither; he had an aura, not of fame, but of greatness. The terrible cancer of the mouth which killed him only eight months later had already attacked him. It was not an easy interview. He was

extraordinarily courteous in a formal, old-fashioned way—
for instance, almost ceremoniously he presented Virginia
with a flower. There was something about him as of a half-
extinct volcano, something sombre, suppressed, reserved.
He gave me the feeling which only a very few people whom
I have met gave me, a feeling of great gentleness, but behind
the gentleness, great strength. The room in which he sat
seemed very light, shining, clean, with a pleasant, open view
through the windows into a garden. His study was almost a
museum, for there were all round him a number of Egyptian
antiquities which he had collected. He spoke about the
Nazis. When Virginia said that we felt some guilt, that
perhaps if we had not won the 1914 war there would have
been no Nazis and no Hitler, he said, no, that was wrong;
Hitler and the Nazis would have come and would have been
much worse if Germany had won the war.

A few days before we visited him I had read the report of
a case in which a man had been charged with stealing books
from Foyle's shop, and among them one of Freud's; the
magistrate fined him and said that he wished he could sentence him to read all Freud's works as a punishment. I told
Freud about this and he was amused and, in a queer way,
also deprecatory about it. His books, he said, had made him
infamous, not famous. A formidable man.

I must return now to the Hogarth Press and its fortunes.
The figures of our income which I have given above on page
142 show that from 1929 to 1939 the Press contributed on
an average £1,100 to our income. During those eleven years
I had to revolutionize the organization of the business. By
1930 we had a clerical staff of three bookkeepers and shorthand typists. We had a representative who travelled our
books: Alice Ritchie was, I think, the first woman to travel
for a publisher and some booksellers did not like the innova-
tion. She was not only a very good traveller, but also a very

good and a serious novelist. The question of the higher command continued to cause us great difficulty and we never found a satisfactory solution. I was determined not to treat publishing as a means of making a living and I was determined not to become a full-time publisher. The choice, therefore, lay between taking a partner or employing a manager competent enough to be responsible for the everyday running of the business so that I could treat it as a part-time occupation.

We started off with the first alternative. From 1924 to 1932 there entered and left the Press a succession of brilliant and not quite so brilliant young men. They entered as managers and potential partners. As I have already said, I think we were trying to get the best of two contradictory worlds and were asking these brilliant young men to perform an impossible feat, namely to publish best-sellers with the greatest professional efficiency for an amateur publisher in a basement kitchen. I still think that the technical efficiency of the Hogarth Press in the years 1924 to 1939 was extremely high, much higher in matters of importance than that of many—I almost wrote most—large and small publishing firms. For instance, for all those years, when we were publishing from 20 to 30 books a year, selling in the first six months of their existence anything from 150 to 30,000 copies, we practically never—I think that I could truthfully say never, but no one would believe me—ran out of bound copies and were unable to supply an order. This was the result of meticulous daily, sometimes hourly, supervision, checking, organization. Professional publishers will probably politely disbelieve me or at any rate will say (privately) that such a record is impossible for any large publishing business. I am sure that they are mistaken. In a previous volume of my autobiography (*Growing*, pp. 107-109) I related my experience with regard to business organization in the government offices of Ceylon. Every head clerk in every kachcheri to which I was appointed, when on my first day in the office, I told him that 'every letter received in this kachcheri after this week must be answered on the day of its receipt unless it is waiting for an order from me or from the G.A.', threw up his hands in horror and despair, and said that this was much too big a kachcheri and received daily far too many letters to make this possible. The head clerk was mistaken and after six months he had to admit it. Whether ten letters or 100 or 1,000 are received daily, they can all be answered on the day of receipt, provided that there is an automatic routine and meticulous checking. It is the same with books. For the efficiency of publishing it is most important that the decision to reprint or rebind for every book on the list should be made at the right moment and for the right quantity so that there is always in stock an adequate number of bound copies to supply the demand (while, at the same time, the publisher, for his own sake, does not print or bind more copies than he can sell). This may seem a fatuously simple and self-evident truism. But a very little knowledge of what goes on behind the scenes will prove how often in practice the truism is neglected. But in 99 cases out of a hundred in which a bookseller has to be told that a book cannot be supplied because it is 'printing' or 'binding', the failure was avoidable and was due to bad organization and slovenly supervision. I should perhaps add that all this, including the truism, often overlooked even by publishers, is that a publisher cannot have an efficient publishing business unless the business has an efficient publisher.

The Hogarth Press was from 1924 to 1939, I repeat, an extremely efficient publishing business, though its methods were in most ways unorthodox. The business side during those years was managed by me and the succession of young men. The first young man to enter and leave the basement in

Tavistock Square—and by no means the least brilliant—was G. W. H. Rylands, universally known as Dadie. We were aiming very high when we took Dadie into the Press and began to turn him into a publisher. Not unnaturally he did not stay long in the basement. When he came to us he was only 22; a scholar of Eton and King's, he had just taken a degree and had written a fellowship dissertation. It was from the first understood that, if he got his fellowship at King's, we should lose him, for he would return to Cambridge to become a don. We treated him rather badly, for almost at once we went off for a week or two, leaving him alone in charge of the Press and of a strange, elderly shorthand typist, who, seeing that he was out of his depth and not very happy, tried to cheer him up by feeding him on sandwiches. Dadie, of course, got his fellowship and, much to our regret, left us for a distinguished career in the university, the arts, and the theatre. We published two books of poetry by him, *Russet and Taffeta* (1925) and *Poems* (1931), and *Words and Poetry* (1927), a very original book of literary criticism based on the dissertation which won him his fellowship.

Dadie was followed by Angus Davidson, who stayed with us from 1924 to 1929; he is now very well known as a translator of Italian books. He was followed by a very young man, Richard Kennedy, a nephew of Kennedy, the architect. Richard was too young to be a manager, and we now had, in addition to the clerical staff, a general manager. In 1931 he left us and John Lehmann entered the Press. His appearance on the scene was to have a considerable effect upon the Press and its fortunes. Unlike the other young men, he took publishing very seriously and became a highly efficient professional publisher. His first term of work with us was short and not very successful, largely owing to my fault. Poor John, like Dadie a product of Eton and Cambridge, only 24 years old when he came to us, was put into a small, dark, basement room, from which he was expected, under my supervision, to 'manage' the publication of 22 books to be published by us in the spring of 1931. These 22 books included Vita's *All Passion Spent* (selling 14,000 copies in the first six months), Virginia's *The Waves* (selling 10,000), a first novel *Saturday Night at the Greyhound* by John Hampson (selling 3,000), and two masterpieces by Rainer Maria Rilke, very difficult publishing propositions 30 years ago, the *Duino Elegies* and *The Notebook of Malte Laurids Brigge*. We published these 22 books—and another 18 in the autumn season of 1931—with a staff consisting of one traveller and four or five in the office. John, Virginia, and I, as well as the 'staff', were expected to be able to take a hand at any and everything, including packing—indeed, we all became expert packers. To pack and despatch 4,000 or 5,000 copies of a book before publication, as we did in the case of *All Passion Spent* or *The Waves*, is a very formidable business, but I do not think that in those days we ever failed to have all our books delivered on subscription orders in the shops well before the publication date. On the top of this John began to help us with the printing. I am not a person who bears even wise men gladly (sometimes and in some ways I bear fools more gladly), and it is not surprising that John soon found the Hogarth Press too much of a good or a bad thing, and left us in September 1932. He returned to us six years later and was Partner and General Manager in the Press until 1946. But that is part of a different story, a different world and phase of my life from that of 1930; it cannot be dealt with here, for it is part of the story of the war years.

John was a young man of 24 when he came to the Press in 1931; I was 51 and Virginia 49. We were well aware that the worst menace of middle age is emotional and mental sclerosis which makes one insensitive to anything in any generation later than one's own. Both as publishers and private persons we wanted, if possible, to keep in touch with the younger generation or generations. In taking John into the Press we had had great hopes that he would help us to do this. A scholar of Eton and Trinity College, Cambridge, a poet himself and the friend of Virginia's nephew, Julian Bell, and the younger generation of Cambridge poets and writers, he seemed to have all the qualities and contacts which we were looking for. In this respect he did not disappoint us and in the two years during which he was with us he helped us to bring into the Hogarth Press some of the best writers of his generation.

The younger generation of the late 1920s and early 1930s was remarkable in quality and quantity. As is usual with the young who have something of their own to say, they were in revolt against their fathers (and mothers, uncles, and aunts). In *New Signatures*, which we published in 1932 and which was and still is regarded as that generation's manifesto,[1] Stephen Spender wrote in his poem 'Oh Young Men':

> *Oh young men oh comrades*
> *it is too late now to stay in those houses*
> *your fathers built where they built you to build to breed*
> *money on money . . .*
> *Oh comrades step beautifully from the solid wall*
> *advance to rebuild, and sleep with friend on hill*
> *advance to rebel . . .*

[1] 'The little book was like a searchlight switched on to reveal that, without anyone noticing it, a group of skirmishers had been creeping up in a concerted movement of attack' (*The Whispering Gallery* by John Lehmann, p. 182). 'The slim blue volume, No. 24 in the Hogarth Living Poets Series, which had caused so much fuss to assemble, came out in the spring of 1932, created a mild sensation then, and has since been taken to mark the beginning, the formal opening, of the poetic movement of the 1930s' (*Journey to the Frontier* by Peter Stansky and William Abrahams, p. 77).

The rebellious youth of the 1930s were nearly all poets, and their knock on the door in *New Signatures* was poetic. Of the nine poets, five were at Cambridge, Richard Eberhart, William Empson, Julian Bell, A. S. J. Tessimond, and John Lehmann; three at Oxford, W. H. Auden, C. Day Lewis, and Stephen Spender; the ninth was William Plomer. Like most young poets, they took a pretty black view of the past, the present, and the future; it might be said of them, to quote Empson, that they 'learnt a style from a despair' (though every now and again, particularly with Julian, in the best tradition cheerfulness would break through). Despair has always been the occupational disease of young poets, but the nine poets of *New Signatures*, it must be admitted, had more reason than most for gloom and foreboding. They looked out upon a world which had been first devastated by war and was now being economically devastated by peace. What could be more grey and grim than the dole and unemployment; the communism of Stalin, the purges, the kulaks, and the Iron Curtain; the tawdry fascism of Mussolini and the murderous nastiness of Hitler and his Nazi thugs? And when they looked from the present toward the future, they could only look forward to dying in another inevitable and futile war. 'Who live under the shadow of a war,' sang Stephen, 'What can I do that matters?' 'A cold wind blows round the corners of the world,' sang William Plomer; 'it blew upon the corpse of a young man, Lying in the street with his head in the gutter Where he fell shot by a revolutionary sniper.... He was rash enough to go out for a breath of fresh air.... Shot through the stomach he took time to die.'

It was John who brought *New Signatures* to us. Through his sister Rosamond he had got to know Stephen Spender and the Oxford poets. Then through his first book of poems, *A Garden Revisited*, which we had published in 1931, he got to know Michael Roberts, who had been a scholar at

Trinity senior to John. He and John devised *New Signatures* and Roberts edited it. In fact, before John came to the Press, we had already published a good deal of work by the rebellious poets of *New Signatures*. Five years before, in 1926, we had published William Plomer's remarkable first novel, *Turbott Wolfe*. In 1929 and 1930 we had published two anthologies, *Cambridge Poetry 1929* and *Cambridge Poetry 1930*. They were anthologies of poetry written, selected, and edited by Cambridge undergraduates. Four of the five Cambridge poets of *New Signatures*—Julian, John, Eberhart, and Empson—were included in these two volumes. We had even penetrated into Oxford, for already in 1929 we had published Cecil Day Lewis's first book of poems, *Transitional Poem*. However, it was not only the emergent poets whom John helped us to keep in touch with. It was through him, via Stephen Spender, that one of the most remarkable and strange of the emergent novelists came to the Hogarth Press, Christopher Isherwood. In 1932 we published *The Memorial*, in some ways his best novel, and in 1935 the brilliant *Mr Norris Changes Trains*.

After John left us and until he returned to us in 1938, I tried the other alternative. We made no further attempt to find a partner, I ran the Press by myself in my own way, with a woman manager. In many ways this worked very well, but it had, from our point of view, great disadvantages. It meant a tremendous amount of work and much more time devoted to the business of publishing than I wanted to devote to it. It also meant that we were closely bound to the business in Tavistock Square and it was impossible to get right away from it for any length of time. Virginia felt this tie to be very irksome, and we were always on the point of throwing the whole thing up or of trying once more to find the ideal partner. But we drifted on, as one does, when something which one has oneself started in life without much thought of the future or of the consequences takes control of one. That did not mean that we let the actual business of the Press drift; we took it very seriously and energetically, and we continued to publish a considerable number of books of every sort and kind. For instance, in 1935, three years after John left us, we published 25 books. These included *Mr Norris Changes Trains* by Christopher Isherwood, *Grammar of Love* by Ivan Bunin, *An Autobiographical Study* by Freud, *Requiem* by Rilke, *A Time to Dance* and *Collected Poems* by C. Day Lewis, together with a great variety of books on politics, education, literature, biography, and travel.

I will leave the Hogarth Press now in its basement in Tavistock Square in our own hands and with us in its hands until the year 1938. It had, as I have said, materially added to our income. We were now very comfortably off, for in the ten years from 1930 to 1939 our average annual income was over three times what it had been when we came to Tavistock Square in 1924. And, as I have also said before, we did not alter the framework of our lives. We lived in the same houses—in Tavistock Square and Rodmell—with the same servants and in the same 'style'. This is shown by the fact that what we spent on that day-to-day framework of living, on house, food, servants, etc., hardly altered—for instance, whereas in 1924 our income was £1,047 and our expenditure £826, in 1934 our income was £3,615 and our expenditure £1,192.

Yet there was one thing which, as Virginia often remarks at the time in her diary, had a great and immediate effect upon the quality and tempo of our life, and the change was directly due to our being able to spend money. In July 1927 we bought a second-hand Singer car for £275. I suppose that the nineteenth-century scientific revolution—in particular electricity and the internal combustion engine—have changed the world—one will probably soon have to say the universe—much more profoundly than anything else which has happened since God rather foolishly said: 'Let there be light'. That God's remark about the light has been the primal cause of infinitely more evil and misery than of good and happiness is certainly true. When one thinks of the two world wars and the destruction of Hiroshima, or even when one compares the car-infested streets of London in 1966 with its humanly inhabited streets which I remember in 1886, one has to say the same about science and the inventions which have given us a Singer car and an aeroplane. Certainly nothing ever changed so profoundly my material existence, the mechanism and range of my every-day life, as the possession of a motor car. Even as an individual, I sometimes think how pleasant the tempo of life and movement was when the speed limit was about eight miles an hour, and I curse the day when I acquired a licence to drive motor vehicles of all groups. But those moments of nostalgic pessimism are rare and unreasonable. There is no doubt whatever that, as an individual, purely as an individual, I have enormously increased the scope and pleasures of living by the six cars which I have owned and driven in the last 40 years.

The most important change and the greatest pleasure came, and still comes, from a holiday 'touring' on the Continent—I do not think that anything gave Virginia more pleasure than this. She had a passion for travelling, and travel had a curious and deep effect upon her. When she was abroad, she fell into a strange state of passive alertness. She allowed all these foreign sounds and sights to stream through her mind; I used to say rather like a whale lets the seawater stream through its mouth, straining from it for its use the edible flora and fauna of the seas. Virginia strained off and stored in her mind those sounds and sights, echoes and visions, which months afterwards would become food for her imagination and her art. This and the mere mechanism and kaleidoscope of travel gave her intense pleasure, a mixture of exhilaration and relaxation.

Before 1928 and the Singer car we used to go abroad, travelling as English people had done for the last hundred years, comparatively slowly, by boat and rail to some place where one would probably settle down for a week or two. Like so much of life before the motor car and the aeroplane, the tempo was slow. One got into the train at Victoria in the afternoon, crossed Paris from the Gare du Nord to the Gare de Lyon in the evening, and woke up to the entrancing moment when one saw the dawn over the Rhône valley and ate one's first *petit déjeuner*. A few years after the end of the 1914 war, when it again became possible to visit France as a civilian, we saw once more out of the carriage window the dawn break over the Rhône valley, for we were on our way to Cassis. In those days Cassis was a small fishing-village between Marseille and Toulon; it had one small hotel, the Cendrillon, at which we stayed. It is almost impossible to believe today that places like Cassis really did exist 40 years ago on any European coast, warmed by the sun and looking upon the blue waters of the Mediterranean or other oarless sea. For it was a pretty, quiet village lying along a restful bay. There was a small resident colony of multi-racial artists, the ominous harbingers of civilization who could already be found infecting the Mediterranean from Almeria to Rapallo. You would usually find three or four visitors, most probably English, at the Cendrillon. Otherwise Cassis still belonged to the people of Cassis. Looking back upon it today, its chief characteristic seems to have been its quiet. It was indeed so quiet that men might have risen up at the voice of a bird. The voice of the motor car was rarely heard in it. People sat in the café and talked or were silent for many hours; or down by the water they leaned against a boat and talked or just looked out across the bay. In the evening the men played boules.

The first time we went to Cassis was, I think, in March. In the mornings we used to go out and sit on the rocks in the sun and read or write. You were alone, the only sounds the water lapping on the rocks or the gulls crying. If you walked up through the wood over the headland to the east, when you came out of the wood on the other side, you had a view of the long line of sandy coast all the way to La Ciotat, Sanary, and Toulon. It was open, flat country, with scarcely a house to be seen until you got to Sanary. Twenty-six years later I was staying in Sanary and visited this stretch of country again, but the country itself had disappeared—it had disappeared under an unending sea of houses and villas, those hideous little villas which the French build all along the Mediterranean coast. I drove from Sanary to Cassis. A stream of perpetual motion, of moving cars nose to tail and tail to nose, ran on both sides of the dusty road between the unending sea of villas. Cassis itself was submerged in cars and villas. Down by the bay the earth was black with human beings; one's ears were deafened by the voice of the loudspeaker and innumerable transistors. The rocks were littered with bottles and paper bags. The scene is the same from Torremolinos to Rapallo, from Ostende to Brest, and from Deal to Land's End.

In the 1920s, as I said, Cassis was a quiet place and, liking quiet, we returned to it. In those days there lived in Cassis one of those curious Englishmen whose complicated characters seem so English and so un-English. Colonel Teed had been Colonel of the Bengal Lancers. When I first met him, I thought: what a perfect Colonel of the Bengal Lancers! A great horseman, the perfect English cavalry officer! I was quite right, but he was also something entirely different, something which one would not have expected to find in the perfect cavalry officer, for beneath the immaculate surface of the colonel of a crack Indian regiment Teed was funda-

mentally an intellectual who liked artists and intellectuals.
He was also a charming man. When he retired from the
Bengal Lancers, he bought Fontcreuse, a vineyard, with a
lovely house a mile or two from the centre of Cassis. There
he made excellent wine. We got to know him, and when
Vanessa followed us to Cassis, she liked the place so much
that she entered into an agreement with Teed which allowed
her to build a villa on his land free of charge and live in it
rent-free for ten or twenty years (I forget the exact figure),
after which the villa would become Teed's property.

Vanessa and Clive and Duncan Grant used to spend much
of their time in the Cassis villa, and this was an added in-
ducement for us to return there. Teed let us have a room in
Fontcreuse and we usually had our meals with the Bells at
the villa. It was a pleasant way of life, so pleasant that at one
moment we began to buy a villa for ourselves near Fontcreuse; it was called a villa, but was in fact a small, rather
tumbled-down whitewashed house. But we did not complete
the purchase. The procedure for an Englishman at that time
to buy a house in Provence was an unending labyrinth;
weary of the interminable business, we began to realize that
our commitments in England, the Hogarth Press, writing,
politics, made it extremely improbable that we should ever
be able to spend much time in Cassis. So the project lapsed.

I am telling this because it explains why we tended, when
we could snatch a few weeks from London and Rodmell, to
make for Cassis. And in March 1928, under the new dis-
pensation, we set out in our old Singer car to drive there. It
opened to one a new way of life. In the old way of travel one
was tied to the railway; as one moved through a country,
one followed the straight steel parallel lines, one had no
contact with the life of the road, the village, and the town.
In Ceylon I had become accustomed to travel freely and
lightly, on a horse or on foot, along empty roads, paths or

game-tracks. I know nothing more exhilarating than starting out in the early morning, just before sunrise, through the jungle or along the straight empty road or the village path for some distant village, an all-day journey with all the day before one. Something of that wonderful feeling of liberation comes—or came—to one when one drove out of Dieppe and saw the long, white, empty road— roads in France, even the *routes nationales*, were in those days empty—stretching before one all the way to the Mediterranean. It is only of this kind of travel, the travel by road, that Montaigne's saying, which I have quoted so often, is really true—it is not the arrival, but the journey which matters.

Our first journey by road on the Continent to the Mediterranean and back was in parts an adventure. It was in 1928 and we crossed from Newhaven to Dieppe on March 26 and for the next six days we rattled along at about 100 miles a day in the old Singer car through Burgundy, and then down the Rhône valley through Vienne, Valence, Montélimar, and so to Provence, through Carpentras and Aix to Cassis. It is the journey, not the arrival, that matters! There are few things in life pleasanter than this long journey south along the white roads through the great avenues of trees and the villages and towns. The slower the better—it is the journey that matters. Even to recall and repeat the names of the towns through which we passed gives me—writing today in a grim grey February day in Sussex—intense pleasure. It is extraordinary how vividly the name of a place will recall to one from years ago the vision of it. I have only to murmur to myself Angunakolapelessa and it brings to me from 50 years ago quite clearly the vision of that small Sinhalese village; I can feel again the whip of heat across my face from the village path; I can hear again the hum of insects across the scrub jungle; I can smell again the acrid smell of smoke and shrubs. So with the names of those towns upon the road to Provence; each recalls the continual change of sound and sights and smells as one journeys south. And there are two particular moments in this journey. The first is at a bend in the road near Montélimar where you suddenly see that the country before you has changed completely—you have left northern Europe and central Europe for the south, you are in Mediterranean Europe. The second is when, after 700 miles of straight Roman roads, your road climbs a hill and from the top of it you see the Mediterranean sea, and at the sight of it I shout, like Xenophon's Greeks: 'Thalassa! Thalassa!'

We stayed for a week in Cassis at Fontcreuse, the Bell family being now established in their villa, and then we set off for Dieppe. I still in those days regarded travel much as I did in my district in Ceylon; I assumed that one could go on in a leisurely way anywhere, knowing one's eventual destination, but not thinking very much of what from day to day lay before one. I therefore mapped out what I thought would be a pleasant route through the centre of France, country which I had never been through. In fact, being fairly ignorant of geography, a subject which as a classical scholar I was never taught at school, I was completely unaware that the centre of France consisted of the massif central, the very formidable mountains of the Cévennes and Auvergne. I was also unaware that in early April one might run into heavy snowstorms on the top of these mountains. So I drove to Tarascon and, after sleeping there, started to drive through Alais and Florac to St Flour and Aurillac. I was rather dismayed later on in the morning to find myself confronted by a formidable, black mountain up which the road began to climb. We climbed and climbed; it grew blacker and blacker both on the earth and in the sky. The clouds seemed to descend upon our heads and we ran into a tremendous snowstorm. The scenery, such as one could see of it, reminded me of Wagner and Covent Garden, those absurdly melodramatic Valkyries' Rocks and Brünnhilde's cave. So Wagnerian was everything that I was hardly surprised when, driving at about 15 miles an hour and peering through the snowstorm, I saw ahead of me suddenly the side of the mountain open into a long tunnel lit by electric lamps. So we passed through the Massif du Cantal, and, when we issued from the tunnel on the other side, it was into bright sunshine. Despite the sun, our troubles had not ended, and my drive from the Cantal to Dieppe was rather a nightmare. I had really only just learned to drive a car and knew very little about either its inside or its outside. I had not realized that the tyres were very worn and the roads very bad. In the 500 miles from the Cantal to Dreux, where at last I succeeded in buying new tyres, we punctured on an average every 25 miles. It seemed to me that there was hardly any road in France on which I had not grovelled in the mud changing wheels.

Such hazards of travel do, however, bring their compensations in curious meetings. On the day we passed through the Cantal in a grim, black, Wagnerian Auvergne hamlet of a few cottages I punctured immediately outside a cottage from which a man came out and offered to repair a tyre. It was raining hard and we went into the cottage and sat talking to the family. A girl of sixteen was sitting at a table writing a letter. The letter was to a 'pen-friend' in Brighton. The girl had never been more than 20 miles from her home, but there is an international organization that organizes these international pen-friends—and there she sat perched on the top of the mountain in the centre of France writing to a girl, whom she had never seen and never would see, who lived ten miles from us in Sussex. She showed us a photograph and the letters of her Sussex pen-friend, the daughter of an omnibus driver in Brighton. In half an hour we were on

terms of warm friendship with the whole family. The internationalism of the savagely nationalistic world of the 1920s
was remarkable.

To drive in a leisurely way through a foreign country,
keeping one's ears and eyes open, is one of the best ways of
getting a vision of both international politics and human
nature. I had a curious example of this in 1935. Vanessa
was in Rome where she had taken a house and studio for six
months. We planned to spend the whole of May abroad, our
idea being to drive through Holland, Germany, and Austria
to Rome and to stay there for ten days or so. In 1935 people
were just beginning to understand something of what Hitler
and the Nazis were doing in Germany. I had only once been
to that country: Virginia and I had stayed for a week in
Berlin with the Nicolsons when Harold was still in the
Diplomatic Service and in the Berlin Embassy. By 1935
Harold had abandoned diplomacy for politics—he was an
M.P.—and journalism, but when I told him my plans for
driving to Italy by way of Germany, he said that he had heard
that the Foreign Office had advised Cecil Kisch of the India
Office that it was inadvisable for Jews to travel in Germany.
He thought it would be as well for me to consult someone in
the F.O. It seemed to me absurd that any Englishman,
whether Jew or Gentile, should hesitate to enter a European
country. I remembered Palmerston's famous speech: '*Civis
Romanus sum* . . . ' and how he mobilized the British fleet
and blockaded Greek ports on behalf of the British subject
Don Pacifico, a Jew born in Gibraltar, in order to recover
£150 damages done to this British subject's house in
Piraeus. Surely, I thought, the British Government in 1935
would insist that the Nazis and Hitler treat an English Jew
as they would any other British subject. However, at that
time I knew Ralph Wigram of the F.O.; he lived in South-
ease, the next village to Rodmell; so when we went down to

Monks House, I rang him up and told him what Harold had said. Wigram said that he would rather not discuss the matter over the telephone and would come round and see me.

When Wigram appeared, I found his attitude rather odd. He said that it was quite true that the F.O. advised Jews not to go to Germany, and officially he had to give me that advice. But privately and as a friend, he could say that he thought it nonsense, and that I should not hesitate to go to Germany. The only thing which I ought to be careful about was not to get mixed up in any Nazi procession or public ceremony. He also gave me a letter to Prince Bismarck, Counsellor at the German Embassy, and advised me to go and see him. So off I went to see Bismarck in the rather oppressive mansion in Carlton House Terrace. Bismarck was extremely affable—of course, my distinguished wife and I must go to Germany. There would be no difficulty of any sort, and he would give me an official letter which would ensure that all government servants would give me assistance if I needed it. He gave me a most impressive document in which Prince Bismarck called upon all German officials to show to the distinguished Englishman, Leonard Woolf, and his distinguished wife, Virginia Woolf, every courtesy and render them any assistance which they might require.

The sequel was amusing, for a marmoset made it quite unnecessary for me to use Bismarck's letter to protect me from the Nazis' anti-Semitism. At that time I had a marmoset called Mitz which accompanied me almost everywhere, sitting on my shoulder or inside my waistcoat. I had acquired her from Victor Rothschild. Victor and Barbara were then living in Cambridge. One hot summer afternoon we drove to Cambridge, dined with them and drove back to London after dinner. We dined in the garden and a rather rickety marmoset which Victor had bought in a junk shop and given to Barbara was hobbling about on the lawn. She climbed up on to my lap and remained with me the whole evening. A month later Victor wrote to me saying that they were going abroad for some time, and, as the marmoset seemed to have taken to me and I to the marmoset, would I look after her while they were away? I agreed and Mitz arrived in Rodmell. She was in very bad condition and I gradually got her fit. She became very fond of me and I of her, and when the Rothschilds returned to Cambridge I refused—much to their relief—to hand Mitz back to them.

Mitz was a curious character. I kept her alive for five years, which was a year longer, the marmoset keeper at the Zoo told me, than the Zoo had ever been able to keep a marmoset. She was eventually killed by a terrible cold snap at Christmas when the electricity failed for the whole of a bitter cold night at Rodmell. During the day she was always with me, but the moment it became dark in the evening she left me, scuttled across the room into a large birdcage which I kept full of scraps of silk. She rolled herself into a ball in the middle of the silk and slept until the next morning—the moment the sun rose, she left the cage and came over to me. She was extremely jealous, a trait which I on occasions took advantage of to outwit her. She was always quite free in the house, but I had to be careful not to let her get out into the garden at Rodmell by herself, for, if she did, she would climb up a tree and refuse to come down. When this happened, I usually succeeded in getting her back by climbing a ladder and holding out to her a butterfly net in which I had put the lid of a tin with a little honey on it. She was so fond of honey that usually she could not resist it and then I caught her in the net.

Late one summer afternoon on a Sunday, when we were just going to get into the car to drive back to London, she escaped into the garden at Rodmell and climbed about 30 foot up a lime-tree at the gate. When I called to her to come down, I could see her small head among the leaves watching me, but she would not budge. I tried the butterfly net trick, but not even the honey would tempt her. So I got Virginia to stand with me under the tree and I kissed her. Mitz came down as fast as she could and jumped on my shoulder chattering with anger. We successfully played the same trick on her another time when she got away into a large fig-tree and I could not dislodge her. She was rather fond of a spaniel which I had at the time and in cold weather liked to snuggle up against the dog in front of a hot fire. She would eat almost anything. Meal-worms and fruit were regular articles of her diet. She once caught and ate a lizard and the Zoo keeper told me that in the open-air cage their marmosets would sometimes catch and eat sparrows. Mitz had a passion for macaroons and tapioca pudding. When given tapioca, she seized it in both hands and stuffed her mouth so full that large blobs of tapioca oozed out at both sides of her face.

I took Mitz with me when on May 1, 1935, we crossed from Harwich to the Hook of Holland. At that time I had a Lanchester 18 car with a Tickford hood so that, by winding the hood back, one could convert it from a closed-in saloon to a completely open car. Most of the day Mitz used to sit on my shoulder, but she would sometimes curl up and go to sleep among the luggage and coats on the back seat. For eight days we toured about all over Holland. Whenever I go to Holland, I feel at once that I have reached the apotheosis of bourgeois society. The food, the comfort, the cleanliness, the kindliness, the sense of age and stability, the curious mixture of beauty and bad taste, the orderliness of everything including even nature and the sea—all this makes one realize that here on the shores of the dyke-controlled Zuider Zee one has found the highest manifestation of the complacent civilization of the middle classes. I have felt something of the same thing in Sweden and Denmark, but I do not think that the Scandinavians have ever reached quite the heights of domestication and complacency attained by the Dutch. It is, of course, easy and, particularly since 1847 and the Communist Manifesto, fashionable to pick holes in bourgeois civilization or savagery, and I am sure that I should soon feel suffocated if I had to live my life in the featherbed civilization of Delft or The Hague. Yet there is in fact a great deal to be said for it and for a short time it is very pleasant to feel that one is in a really civilized country from which nature has been expelled by something more efficacious than a fork. At any rate I prefer the tradition of comfortable civilization in the Netherlands to that of Teutonic sentimental savagery across the border.

Mitz was a great success with the Dutch; wherever we went little groups of people would surround the car and go into ecstasies about 'the dear little creature'. On May 9 we crossed the frontier from Roermond into Germany near Jülich. Immediately I had my first distasteful taste of Nazism. When I went into the Customs office there was a peasant just ahead of me who had a loaded farm-cart. The Customs officer was sitting at a desk and behind him, on the wall was a large portrait of Hitler. The peasant did not take his cap off and the officer worked himself up into a violent tirade against the insolence of a swine who kept his cap on in front of the Führer's image. I do not know whether this exhibition was mainly for my benefit, but I felt with some disquiet that I had passed in a few yards from civilization to savagery, and that perhaps it was just as well that I had Prince Bismarck's letter in my pocket.

The Customs man passed me through without abuse and I drove on to Cologne and from Cologne to Bonn. On the autobahn between these two towns I became more and more uneasy. We seemed to be the only car on the road, and all the way on both sides of it at intervals of 20 yards or so stood a soldier with a rifle. When I reached what must have been more or less the centre of Bonn, I turned a corner and found myself confronted by an excited German policeman who waved me back, shouting that the road was closed to traffic as the Herr Präsident was coming. I tried to find out from him whether there was any road open on which I could drive to Mainz, but he was too excited to do anything but shout that the Herr Präsident was coming.

I turned back and parked the car and we went to see Beethoven's house in order to revive our drooping spirits. Then we had a cup of tea and considered the situation. We were on the right bank of the Rhine and it seemed to me that if the Herr Präsident—whom I wrongly thought was Hitler; he was in fact Goering—was coming to Bonn on that bank, then the road to Mainz on the left bank must be open to traffic. What I had to do was to find a bridge over which I could drive to the left bank. Leaving the tea shop, I stopped a man and asked him how I could do this. He was an extremely kindly German, and he got into the car and guided me across the river. There we were faced by an inexplicable and disturbing sight. On each side of it the main road was lined with uniformed Nazis and at intervals with rows of schoolchildren carrying flags. There were flags everywhere and the singing of Nazi songs. One had to drive extremely slowly as the Nazis were drawn up so as to leave only a narrow strip for traffic. It seemed to me that these regimented crowds were obviously waiting for the Herr Präsident, but, if so, what on earth did it mean? Why were the roads on the right bank closed to traffic so that he could come safely to Bonn by them, if in fact these stormtroopers and school-children were waiting to greet him on the main road, open to traffic, on the left bank?

At any rate, we had run straight into the kind of situation

which Wigram had warned us to avoid. Here we were closely penned in by what, looking down the road ahead, seemed to be an unending procession of enthusiastic Nazis. But we soon found that there was no need for us to worry. It was a very warm day and I was driving with the car open; on my shoulder sat Mitz. I had to drive at about 15 miles an hour. When they saw Mitz, the crowd shrieked with delight. Mile after mile I drove between the two lines of corybantic Germans, and the whole way they shouted 'Heil Hitler! Heil Hitler!' to Mitz and gave her (and secondarily Virginia and me) the Hitler salute with outstretched arm.

This went on, as I have said, for mile after mile, and eventually I could stand it no longer. I decided to turn off the road, go down to the river, and find an hotel where we could stay the night. On the bank of the Rhine, which seems to me one of the few really ugly rivers in the world, at a place called Unckel, we found a very large hotel. We were the only guests, and we had a curious experience there which threw an interesting light on the view which some Germans took of Hitler and the Nazis in 1935. We dined in an immense, long dining-room. We sat at a table at one end and at the other end the proprietor and his wife had their dinner. There was no other diner and a solitary waiter waited upon us. Towards the end of dinner the proprietor came over and asked us whether we had been satisfied. After some desultory conversation, I asked him whether he knew what the explanation was of the Herr Präsident, the closed and open roads, and the lines of expectant Nazis. He immediately shut up and said that he knew nothing about that, but he did not leave us and, after some rambling conversation, asked me where we came from. When I said from Tavistock Square in London, he suddenly changed into a completely different person: he saw that to an Englishman who lived in Tavistock Square it was safe for him to say anything.

His lamentable tale came pouring out of him. He had been a waiter for many years on the banks of the Thames at Richmond. Then he had returned to Germany in order to marry; he himself would have liked to go back to England, but his wife could not speak English, so he became manager of the Unckel hotel. Before the Nazis appeared on the scene, it was a pleasant and prosperous place; it was always full of young people, for students used to come up the river from Bonn and enjoy themselves. A short time before Hitler began to dominate the scene, our melancholy host had been offered the managership of a London hotel, actually in Tavistock Square itself—there were tears in his eyes as he told us this. He wanted to accept, but his wife could not face a great foreign city in which she would not be able to speak a word of the language. So he refused. Then the Nazis came into power and life in Unckel became hell. 'If one says a word of criticism,' he said, 'one is in danger of being beaten up. It is all processions and marching and drilling. And my business is ruined, for the students in Bonn are kept so busy marching and drilling that they hardly ever now come up the river from Bonn. *Und nun bin ich ins Gefängnis* (and now I'm in prison). They will never let me out; it's impossible to get out of this country.' Here the waiter, who had been standing and silently listening, burst out: 'I'm going to get out. It's terrible here—I'm going to get out—Oh, yes, one can—I shall go to America—that's the place to live in.'

Next day we left the manager, his wife, and the waiter in tears, and drove through Mainz and Darmstadt to Heidelberg, and from Heidelberg through Stuttgart and Ulm to Augsburg, and from Augsburg through Munich to the Austrian frontier, and so to Innsbruck, where on May 12 snow was falling. We did not enjoy this; there was something sinister and menacing in the Germany of 1935. There is a crude and savage silliness in the German tradition which, as one drove through the sunny Bavarian countryside, one felt beneath the surface and saw, above it, in the gigantic notices outside the villages informing us that Jews were not wanted.

Not that we had any difficulty anywhere. We forgot about Bismarck's letter, for Mitz carried us through triumphantly in all situations. Pig-tailed schoolchildren, yellow-haired Aryan Fräuleins, blonde blowy Fraus, grim stormtroopers went into ecstasies over *das liebe, kleine Ding*. What was it? Where did it come from? What did it eat? No one ever said a sensible word about Mitz, but, thanks to her, our popularity was immense. In Augsburg in a traffic jam a smiling policeman made cars get out of the way for her and for us. It was obvious to the most anti-Semitic stormtrooper that no one who had on his shoulder such a 'dear little thing' could be a Jew.

As we approached the Austrian frontier, I told Virginia that I proposed to show Bismarck's letter to the Customs officers in order to see what effect it would have. When I drew up at the barrier, the usual scene took place. As soon as the officer saw Mitz on my shoulder, he shouted to his wife and children to come out and see *'das liebe kleine Ding'*. We were soon surrounded by two or three women, four or five children, and several uniformed men. The usual *'Achs!'* and *'Os!'*, the usual gush of gush and imbecile questions. I thought that we would never get away, but eventually they calmed down and passed us through without any examination of anything. At the last moment, shaking hands all round, I presented Prince Bismarck's letter. The effect was instantaneous and quite different from that of Mitz, the marmoset. The chief officer drew himself up, bowed, saluted, clicked his heels together, drew all the uniformed men up in line, and, as we drove away, they all saluted us.

Next day we crossed the Brenner and drove down through Italy to Rome, staying a night at Verona, at Bologna, and at

Perugia. How different in those menacing days was the Fascism of the Italians from the Nazism of the Germans! Beneath the surface of Italian life the vulgar savagery of Mussolini and his thugs who murdered Rosselli was, no doubt, much the same as that of Hitler and Goering; but whereas German history has never allowed civilization to penetrate for any length of time, either widely or deeply, into the German people, Italian history has been civilizing the inhabitants of Italy so deeply and so perpetually for over 2,000 years that no savages, from Alaric and his Germanic hordes to Mussolini and his native Fascists, have ever been able to make the Italians as uncivilized as the Germans. In 1935, therefore, whether Jew or Gentile, you did not require either a marmoset or a Prince Bismarck to protect you from the native savages.

The native savages of Italy delighted in Mitz in the same childish way as those of Holland, Germany, and Austria. It is perhaps interesting, from the anthropological and historical angles, to recall the reactions of the people of Holland, Germany, Austria, Italy, and France to Mitz in 1935. In 25 days I spent in the first four countries dozens of Dutch, Germans, Austrians, and Italians spoke to me about Mitz. They all made one or two of five or six standard remarks or asked one or two of five or six standard questions. The remarks and questions were banal, childish, or sometimes incredibly silly. Mitz was mistaken for practically every existing small animal including a rat and a bat. On Sunday, May 26, I crossed the frontier into France at Ventimiglia and had *déjeuner* at Menton, intending to drive on and spend the night at Aix. The Sunday traffic along the Riviera road was so abominable that after Nice I decided to turn north and drive by the unfrequented minor roads to Draguignan and so to Aix. When we got to Draguignan, we had had enough of it, and we decided to stay there for the night.

The stretch of country which runs north of the Riviera road from Grasse to Draguignan is grey and grim; Draguignan itself and its inhabitants are rather grey, grim, and chilly. The woman in the hotel, when she saw Mitz, gave her a chilly reception and refused to allow me to bring her into the hotel. We had at last, after travelling 2,469 miles on the continent of Europe, reached a country in which a marmoset was not a dear, little thing. I left Mitz for the night locked up in the car on the road in front of the hotel. Next morning, as soon as I got up, I went out to see that she was all right. She was sitting on the steering wheel and a soldier was standing on the pavement watching her through the window. I took her out and fed her and the soldier watched and talked to me about her. He was an ordinary soldier, and he talked about her in an adult, intelligent way—he was the first man, woman, or child who had done so in 2,469 miles and he was also the first Frenchman who had talked to me about Mitz. The reaction of the man in the street to a marmoset sitting on the steering wheel of a car teaches one something, I think, about the intellectual tradition, even the civilization, of the country to which he belongs. There are a good many things which I do not like in the French tradition, but its scepticism and respect for intelligence seem to me admirable.

I must now say something about my writing and my political activities in the years between the wars; the two occupations were closely connected; they were complementary attempts, the one theoretical and the other practical, to understand and to help to solve what seemed to me the most menacing problems left to us by the war in a devastated and distracted world. There were in fact two vast, oecumenical problems which threatened, and still threaten, mankind and are interrelated: first, the prevention of war and the development of international government; secondly, the dissolution of the empires of European states in Asia and Africa which seemed to me inevitable and which would cause as much misery to the world as war unless the Governments of the great imperial powers recognized the inevitability, and deliberately worked for an orderly transference of power to the native populations, educated for self-government by their rulers. I had already begun to think and write about these two questions before 1920. In *International Government* I had shown that in fact the relations of states had for long been regulated or not regulated by a system partly of complete anarchy and partly of rudimentary international government, and I had argued that war could not be prevented in the complicated modern world unless some kind of League of Nations system could be established under which the relations of sovereign states would to some extent be controlled by law and order, by international government. I continued to write about this, editing for instance *The Intelligent Man's Way to Prevent War* in 1933, but, as I shall explain more fully in a moment, between 1920 and 1939 I became engrossed in a much wider and more fundamental subject of which the problem of preventing war was only a part. As regards imperialism, I lectured on this subject and I wrote two things specifically about it: *Imperialism and Civilization* in 1928 and *The League and Abyssinia* in 1936.

But as a writer, during the 20 years between the wars, and even beyond that for another 14 years until 1953, I devoted myself to a single subject, doing a great deal of work upon it and writing three books, *After the Deluge* Vol. I (1931), *After the Deluge* Vol. II (1939), and *Principia Politica* (1953) which was really intended to be the third volume of *After the Deluge*. There are 1,000 pages in these books and, I suppose, about 300,000 words. To all intents and purposes they have been a complete failure, but, though like most people I would rather succeed than fail—whether in a game of chess or bowls or in one's life work—in many ways I do not regret them. No doubt owing to the delusion of parentage, I see in my offspring merits invisible to other people. I still think that the subject of these volumes is of immense importance to the historian, the philosopher, the psychologist, and the politician, and that it has very rarely been posed and faced in the way I tried to pose and face it, and I still think that there are a certain number of words out of the 300,000 written by me which contribute something of truth and importance to the subject. As this is an autobiography of my mind as well as of my body, I propose to say something about this.

A good deal of mythological nonsense has been written about the impact of the 1914 war upon left-wing intellectuals, a pack or sect to which, I suppose, I have always belonged. The myth is that pre-war liberals all over Europe believed in the inevitability of progress and the complete rationality of man, the political animal, and that, as the war destroyed the foundations of their political beliefs, they became completely disorientated and exploded. Most intellectuals, dead or living, about whom I have known anything, have resembled Diogenes and the author of *Ecclesiastes* rather than Mr Chasuble, Tom Pinch, and Mr Micawber rolled into one innocent and imbecile optimist. I, like nearly all of them, have never believed that progress is inevitable or that man is politically rational. I was born, as I have recorded, in 1880 into a comfortable, professional middle-class family in Kensington. We had been affluent while my father was alive; his death made the position of the family for many years economically precarious. But in Kensington and Putney, at St Paul's School and Trinity College, Cambridge, life was lived in an atmosphere of social stability and security. The battle of Waterloo was fought 85 years before I went up to Cambridge; the Crimean war had been fought nearly 30 and the Franco-German war nearly 10 years before I was born. There were some signs that European civilization might develop more widely and quickly on the basis of liberty, equality, and fraternity in the twentieth century. Despite colonial wars and rivalry, and despite the endemic danger to Europe from the confrontation of the Triple and the Dual Alliance, to be moderately optimistic about the social and political future of the world was in 1900 not unreasonable.

Moderate optimism was, I think, the attitude of most intellectuals and intelligent people—not of course synonymous—during my lifetime up to the year 1914. The war was a tremendous shock to intellectuals, as it was to the world and all its inhabitants. When Austria invaded Serbia and Germany Belgium, it was one of the great turning-points in human history. There was no longer any place for 'moderate optimism'; as far as I was concerned, it seemed to me that events had proved that in the modern world war and civilization were incompatible, and that, when the war ended the supreme political problem was to find means, if possible, for preventing war. That meant that one must find the chief causes of war and one must discover methods to destroy or counteract those causes. As in all cases of political action, to do this required the use of reason. To find the causes of social or political phenomena you have to use your reason to analyse a series of complicated situations or events; to find means of influencing or altering the series of events requires a constructive use of reason. To say this does not mean that one believes that human beings always act rationally. If you say to a man: 'If you walk over that precipice, you will fall 300 feet on to a hard rock and almost certainly kill yourself', you are using reason analytically and constructively to explain the truth about a situation, cause and effect and the result of future action, to an individual. You do not imply any belief that *he will* use his reason and so avoid falling over the precipice. The problems of history and society are subject to precisely the same laws of reason and unreason, cause and effect. It is as simple as that, though unfortunately in history, the simplicity is delusive because of the immense complication of historical situations and events, and therefore of their causes and effects. Moreover, individuals are rather more rational about walking over precipices than states and governments.

When I began to consider the history of war in Europe in the light of analytic reason, it seemed to me, as it did to many other people, that war would sooner or later be inevitable unless there was at least a rudimentary system of international law and order which would provide for the peaceful settlement of international disputes. That was the theme of my book *International Government*, in which I drew the outline of the structure and functions of a League of Nations. But I never thought or said that a League of Nations, even an effective League, was the only thing necessary to prevent war. No large conglomeration of civilized human beings has ever been able to exist anywhere in comparative peace and prosperity without a system of law and order, without some kind of government whose power to enforce the law against the individual law-breaker rests ultimately upon some kind of force. To say that does not mean that one believes that one only had to have laws, courts, and police to produce law-abiding individuals and a peaceful and civilized government and state. All this is also true of international society in which the units are independent sovereign states.

Ruminating on these matters I had come to three conclusions: first, that in the twentieth century science and industry made war and civilization incompatible; secondly, that without some kind of League system war would be practically and eventually inevitable; thirdly, that war was part of a much wider social phenomenon and problem, namely the government of human beings and what I call communal psychology. It was this third conclusion which led me into spending the next 20 years in writing three books. My mind led me down this path in the following way. When I wrote *International Government*, I did a good deal of reading and thinking about the wars of the eighteenth and nineteenth centuries and about the development of international relations and international government. The more I read and thought, the more interested I became in certain historical facts.

I found that historians accepted it as a fact that the principalities and powers, the captains and the kings, the governments and statesmen who 'made war' on one another had always done so for certain 'objects', and apparently the common people who fought and died in the armies of these rulers accepted these objects until the moment of either final victory or final defeat. It seemed to me curious that, when one examined and analysed these 'objects' for which wars had been fought in Europe ever since the revolutionary and Napoleonic wars, one found that they consisted apparently of beliefs and desires, and that the objects, the beliefs and desires, for which we fought the 1914 war appeared to be precisely the same as those for which our ancestors had fought in 1815 at Waterloo—indeed, they were not very different from those for which, according to Herodotus, the soldiers and sailors of Sparta and Athens fought against Xerxes at Marathon, Thermopylae, and Salamis 2,400 years ago. The revolutionary armies towards the end of the eighteenth century fought for liberty, equality, and fraternity and went into battle singing the Marseillaise; the Napoleonic army fought its way to the walls of Moscow for the same objects and, according to Schubert's famous song, in defeat struggled back to France through the snows of Russia and Poland still singing the Marseillaise. If you open the pages of Herodotus, written, I repeat, over 2,000 years before, and read the wonderful speech in which the Greek unsuccessfully tries to explain to Xerxes—sitting on the shores of the Hellespont and reviewing his mighty fleet—tries unsuccessfully to explain to him what it means to be a Greek and a free man and why one Greek will beat three Persians because he is fighting for—you will find that to all intents and purposes the Greek is saying that he is fighting for liberty, equality, and fraternity. And when in August 1914 the British Government, through the voices of the Prime Minister, Henry Asquith, and the Foreign Secretary, Sir Edward Grey, good liberals echoing the voices of Themistocles and Leonidas, Mirabeau and Lafayette, told us that we were fighting to protect the rights of small nations, the freedom of Serbia and Belgium, from aggression by the Great Powers Germany and Austria, were not statesmen and history once more proclaiming that governments and individuals were fighting a war for liberty, equality, and fraternity?

It seemed to me that, if one wanted to understand the causes of war and perhaps discover some means of preventing it, one must investigate more closely these statements of historians, generals, and prime ministers that wars are fought by nations for objects like liberty, equality, and fraternity. If the statements are true, wars are fought for beliefs and desires, for, if you are called upon and agree to fight for liberty, it means that you believe liberty to be so good and you desire it so deeply that you are prepared to fight and die for it. Moreover, if these beliefs and desires really do influence or cause communal actions like war, then they can accurately be described as communal beliefs and desires. If in 1914 the British answered Kitchener's appeal and joined up because they believed Asquith's appeal to fight for liberty, then it is true that there was a general acceptance in the community of the belief about and the desire for liberty and also a general agreement that these communal beliefs and desires should lead to communal action in war.

Such communal beliefs and desires connected with war seemed to me part of something much wider which I called communal psychology. For according to historians and statesmen communal action is almost always influenced or determined by communal beliefs and desires. For instance, they would hold that the communal belief that political equality was desirable and that it was unattainable unless every adult male had a vote determined the communal action in Britain which produced the Reform Acts of 1832, 1867, and 1884. Again, according to history, the political and economic beliefs and desires of a German Jew enunciated in his book *Das Kapital* in 1867 and in the Communist Manifesto became 50 years later the communal beliefs and desires which caused the Bolshevik revolution and has subsequently led to the spread of communism and the emergence of communist governments all over the world. Again, the paranoid beliefs and desires of a disgruntled Austrian corporal about Germans and Jews became the communal beliefs and desires of Nazism and the Third Reich which produced the second great war and the murder of over five million Jews in concentration camps and gas chambers.

I thought, and still think, that there are some very important and very strange things implied in these beliefs of historians and politicians regarding communal psychology. In every case they imply that some social end or situation is desirable or not desirable and that it can be attained or prevented by certain communal action. If this be true, it means that communal action or the events of history can be and are determined partly by emotional judgments of value —this or that is desirable or not desirable—and partly by reason, the convincing of large numbers of people that something, which is communally desirable, can only be attained by some specific communal action.

Though this is always assumed to be true both in theory or in practice—whether in Thucydides's history of the Peloponnesian war, Gibbon's *Decline and Fall*, a meeting of the Cabinet in 10 Downing Street, or of the Praesidium in the Kremlin—there are practically no serious studies by sociologists, psychologists, or historians of how this process of interaction between communal beliefs and desires and communal actions works or indeed of whether in fact it does work. I decided that, though I was not a professional sociologist, historian, or psychologist, I would try to do as an amateur what the professionals had left undone. In 1920 I had finished and published my book *Empire and Commerce in Africa* and in that year I began to read for and study this problem of communal psychology. Though the subject was vast and complicated, my object could be stated clearly and simply. I proposed to study intensively the years 1789 to 1914 and to try to discover what the relation between the communal beliefs and desires regarding liberty, equality, and fraternity and communal action had been during those years, i.e. what, if any, had been the effect of those communal beliefs and desires not merely upon war and peace but upon historical events generally. I had no idea in 1920 that I should be working and writing on this subject for the next 33 years.

I worked at it, it is true, continually for those 33 years, but I had so many other occupations that I spent a good deal less than half my time upon these books. It was not until 1931 that I published the first volume of *After the Deluge*; it was concerned with democracy and democratic psychology as they developed in the eighteenth century and in the American and French revolutions. Eight years later I published *After the Deluge*, Vol. II, which dealt with the communal psychology of the years 1830-1832, the effect of the communal beliefs and desires regarding liberty, equality, and fraternity on the historical events of those years. 1939, year of its publication, brought down upon Europe and the world the second great war. The years in which the first two volumes had been published had turned out to be between the deluges. I intended to write a third volume, but obviously it could not be entitled *After the Deluge*, Vol. III. It was Maynard Keynes who said to me that what I was really trying to do in these volumes was to analyse the principles of politics and I ought to call the third volume *Principia Politica*. I followed his advice, but it took me a long time to write the book, and *Principia Politica* was not published until 1953.

These three volumes have been, as I have said, to all intents and purposes a failure. The first of the three was published as a Penguin and I occasionally still get a letter of appreciation from some unknown person which faintly stirs in me the pleasure of authorship. All three on the whole had an unfavourable press, and *Principia Politica* was received with derision by the Oxford professional historians who do so much reviewing and who rapped me over the knuckles for having the effrontery to be a member of 'Bloomsbury' and use a title which recalled the works of Newton, Bertrand Russell, and G. E. Moore.

I find it quite difficult to be certain in my own mind as to what 'in the bottom of my heart' is my real attitude towards my own books and criticism of them. My disappointment at the reception and fate of these books has been fairly deep, though not very prolonged. No doubt I was considerably annoyed by reviews which dealt at length with 'Bloomsbury' and the title of the book but never so much as mentioned what the book was attempting to do. No one likes to spend

23 years of his life and nearly 300,000 words on something which is invisible to a Fellow of All Souls or Magdalen. I still think that what I was attempting to do is of immense importance to history and sociology, and that professional historians and sociologists have never attempted to do it.

I am a highly prejudiced reader of these three volumes, but I still think that there is a little more in them than was seen by the professional historians who reviewed them. In 1953 at the age of 73, however, I had to make up my mind whether I should carry on with my original intention and plan to write a fourth and even a fifth volume. I decided not to do so. The three volumes were a failure and I was not prepared to spend another five or ten years and another 200,000 words with the same result. I was, as I say, disappointed, but I do not think that the hurt went very deep or was prolonged. It is interesting, I think, to observe the attitude of different writers (and of oneself) to their writings. I have produced about 20 books, and, like most writers, I probably have, in the pit of my stomach, a better opinion of them than other people. But I am not really much concerned about them and people's opinion of them, once they have been published, and I do not feel the slightest interest in their fate after my death. It is here that I find the attitude of so many writers, good or bad, very strange. They seem to regard the fate of their books as if it were the fate of themselves and they seem to see in the book shops and libraries an unending struggle between mortality and immortality. I could never quite understand Virginia's feeling about her books and their reputation in the world. She seemed to feel their fate to be almost physically and mentally part of her fate. I do not think that she had any belief in life after death, but she appeared to feel that somehow or other she was involved in their life after her death. Being so intimately a part of herself, a hurt to them was felt as a hurt to her, and her mortality or immortality was a part of their mortality or immortality.

The crux of the matter is perhaps there, in the immortality of the soul! The difference in the attitude of different writers to their work probably depends upon whether or not deep within them, even unconsciously, they still believe that they may be immortal. I suspect that Virginia, though she did not believe in life after death, did believe in her life after death in *The Waves*, and not merely in the life of *The Waves* after her death. Even if I had written *The Waves* or *Hamlet*, I do not think I could possibly have felt like that. I cannot believe that death is anything but complete personal annihilation. I cannot, therefore, feel any *personal* interest or involvement in anything of mine after I have been annihilated. I should like to know what happens on the day after my death, e.g. what horse wins the Derby if I die the day before Derby day, and I should like to know what happens to my books after my death, but as I shall never know either, annihilation makes it all one for me. The fate of my books, even before my death, loses some of its importance for me, and this in turn diminishes both the pleasure in success or the pain of failure.

And now I must leave books and theory, and say something of the practice of politics. As a practical politician, I worked mainly in the Labour Party and the Fabian Society, but before I deal with that side of my life, I must mention two other political activities. The first was journalistic. In 1930 I helped to start the *Political Quarterly*. The main credit for successfully launching this journal and for its still being successfully published today, 36 years later, must go to Professor William A. Robson. Willie Robson's enthusiasm and pertinacity succeeded in getting enough capital to start the journal—by no means an easy task—and an admirable group of writers to write for it. He and Kingsley Martin were the first joint-editors, and the Editorial Board consisted of A. M. Carr-Saunders, T. E. Gregory, Harold Laski, Maynard Keynes, Sir Arthur (later Lord) Salter, Sir E. D. (later Lord) Simon, and myself. It was, as the names show, Left Wing politically, but of irreproachable respectability. It proclaimed its object as 'to discuss social and political questions from a progressive point of view; to act as a clearing-house of ideas and a medium of constructive thought'. Its standard was from the beginning extremely high and has remained so for 36 years. A journal of this kind cannot be popular; it is written very largely by experts for an elite, for Members of Parliament and civil servants in the arena of practical politics, and in the academic arena by experts for experts in sociology, politics, law, and history. It can only succeed, indeed it can only justify its existence, by providing ideas for or influencing the ideas of a comparatively small number of 'men at the top'.

When Kingsley became editor of the *New Statesman*, I took his place as joint editor with Willie Robson on the *Political Quarterly*, and for some time during the war I was sole editor and even to some extent its publisher. I continued to be an editor until 1959 and remained on as literary editor until 1962. Having been for over 30 years editorially responsible for it, I am necessarily not an unprejudiced witness regarding its merits and influence. Practically all journalists, from the great Press Lords down to the humblest reporter, suffer from the grossest delusions about the 'influence' of the newspaper which they own, edit, or write for. The megalomania probably increases as you go up from the humble reporter through the pompous editor to the paranoiac owner. I am not sure that the evidence, such as it is, does not point to the fact that the larger the circulation of a paper, the less influence it has upon the opinions of its readers, or even that the influence of every paper is in inverse proportion to the number of copies sold or to the number of people who buy or read it. Certainly the millions who read the popular press seem to be singularly impervious to its propaganda, particularly political propaganda, for vast numbers must vote Labour who habitually read anti-Labour dailies. I do not know whether this should be a subject for rejoicing among the angels in heaven or the Left Wing intellectuals on earth. It depends upon what is the explanation of this curious phenomenon. If the millions who read the popular dailies are interested only in sport, battle, murder, and sex, and therefore are inoculated against the opinions of the proprietor and editor, there is not much reason for rejoicing; but if large numbers have learnt to doubt whether the North-cliffes and Beaverbrooks of the twentieth century are good political advisers to follow, I should feel faintly encouraged.

I feel some encouragement too when I contemplate the other end of the scale where are journals of small circulation like the *Political Quarterly*. There is no doubt that, if their standards, both journalistic and intellectual, are high, they can have considerable influence. The reason is that they are, as I have said, written by experts for experts, or, from another point of view, they are professional or trade papers. The *Political Quarterly* is partly a technical paper in which the professional politician, the administrator, or the civil servant can find information and ideas of the greatest importance to his work and unobtainable elsewhere. No one reads the popular dailies for serious ideas, and practically no one takes their ideas seriously; their main function is entertainment, to titillate the universal desire for sex, violence, gambling, and the royal family. No one reads an article by Professor Robson on local government or by Sir Sydney Caine on the common market for entertainment or for any form of sexual or monarchical titillation, but they may be absorbed and fascinated by the ideas. That is why we were able to start the journal with a capital of a few hundred pounds and we have been able to go on for 36 years without having to raise any more capital. It is perhaps also the reason why its influence has been much larger than its circulation.

I have said that I dislike journalism and editorship, because as soon as one has finished off one number, one has to begin to think about the next. This applied to the editorship of the *Political Quarterly*, but as there were three months instead of seven days, as with the *Nation* and *New Statesman*, between each issue, the perpetual turning of the wheel was longer and slower. On the whole I enjoyed my work on it, because I was extremely interested in the ideas with which it dealt and was working with people pursuing the same objects. I also enjoyed the second by-product of my political activities. In 1938 I was appointed a Member of the National Whitley Council for Administrative and Legal Departments of the Civil Service, and I remained a member for the next 17 years.

The nature of the work of the Whitley Council for the Civil Service is not widely known. It consisted, so far as I was concerned, of sitting three or four times a year on arbitration cases. I found the work extremely interesting and even sometimes amusing, and also exasperating. The British Civil Service does not strike, and an elaborate procedure has been devised in the Whitley Council system for the peaceful settlement of economic disputes, i.e. for determining scales of payment and conditions of employment without resort to strikes or lock-outs. In 1938 the staff side of the Civil Service was organized in the Civil Service Clerical Association, which was in fact the civil servant's trade union. Any claim by a class or section of the Service for higher pay or improved conditions was made to the Treasury through the Clerical Association and negotiations then began between the Treasury and the Association. If agreement was reached, well and good, but if, after a time, the two sides could not agree, the claim was remitted to arbitration. The arbitration tribunal consisted of a permanent chairman, who was a lawyer appointed by the government, and two other arbitrators drawn from two panels. One panel was appointed by the Treasury, and consisted mainly of Directors of banks or railway companies; the other consisted of persons nominated by the staff side, by the Clerical Association. The Association had asked me to agree to serve and had nominated me for their panel.

As Assistant Government Agent in the Hambantota District of Ceylon for nearly three years, I had had a fair amount of judicial experience, for as Police Magistrate and District Judge I had had to try both criminal and civil cases. To try any kind of case in a judicial capacity I find extraordinarily fascinating. The fascination to me consists largely in the curiously complicated state of mind into which you, as a judge, have to get, if you are to be a good judge. Your mind, like Caesar's Gaul, has to be divided into three parts, and yet, like Gaul, maintain its unity. First, your mind must work intellectually with great quickness and concentration upon the facts, for the first essential is that the judge should understand and interpret the facts, which are often connected with spheres of life and activities of which the judge has no previous experience. (In Civil Service arbitration cases one was continually having to understand and interpret very complicated facts about the details of work or occupations of which one was quite ignorant before one began to try the case.) I thoroughly enjoy this kind of intellectual problem. Secondly, no one can be a good judge unless he can combine, with this quick intellectual understanding of facts, an intuitive sensitiveness to human witnesses and their evidence. Often it is only by hearing and, as it were, feeling a witness that you can accurately interpret and assess the value of his evidence.[1] The third requisite of a good judge is, perhaps, in some ways the most difficult—it is complete and unfailing impartiality. Complacent prejudice is the occupational disease of judges. It can make the judge incapable of understanding and interpreting the facts or of judging the character of a witness and the value of his evidence. On the bench one has to be perpetually on one's guard against oneself, to prevent one's previous beliefs and prejudices interfering with one's acceptance or rejection of facts and arguments. But still more necessary is it consciously to watch and thwart one's own instinctive prejudices for and against persons. A woman enters the box oozing feminine charm—how difficult it is to regard her and her evidence exactly as one does the next witness who fills one with physical repulsion. And it requires an even more ruthless act of will against oneself to force oneself to judge with complete impartiality the value of the evidence of a man whose appearance you dislike.

[1] This is well understood and theoretically admitted in British Courts of Appeal, though not always honoured in practice by Appeal Court judges. In Ceylon I learned by experience that, where there was a direct conflict of evidence, often suddenly some small thing, almost impossible to describe accurately—a gesture or movement of the witness perhaps—would reveal to one where the truth lay. Of course, one may have deluded oneself, though there were cases in which I would have staked my life on the accuracy of my judgment. The Appeal Court judge did not always agree. I still remember after 50 years a case in which a villager suddenly appeared in the kachcheri hauling along a man and a cow, claiming loudly and passionately that it was his cow and that the man had stolen it. There was a complete conflict of evidence and suddenly one of the witnesses said something in such a way that I was absolutely convinced that the cow had been stolen. I gave my verdict accordingly and it was set aside in appeal. I am certain that thus the thief obtained the cow, and after my death, if I find myself in heaven, the first thing I shall do is to ask Saint Peter whether my judgment was correct. I am sure that he will answer: 'Yes'.

This last point is to me so interesting that I cannot resist saying a word or two more about it. The behaviour of Lord Hewart in the Court of Appeal, described by me on page 137, is a good example of judicial injustice caused, in part, by the judge allowing his prejudice against a man's appearance to make him give a grossly inequitable judgment. I once watched an even worse exhibition of prejudice by Mr Justice Avory at the Old Bailey. I had been summoned as a juror and was waiting in court for the next case while Avory tried a working-class woman for stealing a piece of luggage at Victoria Station. There was no doubt about the facts. She had been convicted several times of the same offence: she loitered on a railway station platform and walked off with the most likely looking suitcase of some first-class passenger. There was nothing against her character except this inveterate habit of pinching first-class passengers' luggage. She was not attractive to look at, but I was convinced by everything she did and said that, apart from this habit, she was an exceptionally nice person. If ever there was a case of a criminal whose crime was in the eyes of God and medical science not a crime, but a symptom of mental or emotional disorder which might well be curable, there it stood in the dock facing us all in the dingy court, including Mr Justice Avory and the jury. I say that it might well have been curable because I happened to have had personal knowledge of a similar case in a very different walk of life from that of the North Country working-class woman being tried by Avory. I knew a young man, born into a highly respectable professional family, himself in a first-class professional post after a successful university career. Suddenly he was arrested for precisely the same offence as this woman, and it was discovered that he had habitually for some time stolen luggage at railway stations. Like her, normal and law abiding in every other way, he had this uncontrollable compulsion to steal suitcases from railway station platforms. He had the good fortune to be tried by an intelligent and sympathetic judge and he was discharged on condition that he took psychiatric treatment. He did so and was completely cured, ending his life as a respected member of a learned profession.

Very different was the treatment which the working-class woman received from Mr Justice Avory. The indignation which I felt while I watched and listened to him as he summed up rises again in me today. He was an exquisitely elegant man in his wig and gown and immaculate lace, and behind the icy ruthlessness of his pitiless summing up against her I felt the righteous indignation of the first-class passenger confronting the thief who had stolen his suitcase. The jury rightly found her guilty and Avory gave her the maximum sentence. Justice had been done, but—

> A man may see how this world goes with no eyes. Look with thine ears: see how yond justice rails upon yond simple thief. Hark in thine ear: change places; and, handy-dandy, which is the justice, which is the thief?

But I must return to the Civil Service Arbitration Tribunal. There, of course, there was no 'simple thief' before us, but, handy-dandy, much the same attitude was required, if one was to be a good arbitrator, as was necessary if one was to be a good High Court judge in the Old Bailey. The procedure was that, when one was chosen from the panel to try some case, one received the detailed claim by the Association or, in some cases, a trade union, and the answer by the Treasury, both in writing. Then on the day fixed for the hearing the three arbitrators heard both sides and any witnesses that they wished to call. When I first became an arbitrator, the case for the staff side was almost always conducted by W. J. Brown, Secretary of the Civil Service Clerical Association. A very able and rather flamboyant man, he was a first-class advocate, and, as he had considerable experience and was meticulous in mastering every detail in the most complicated claim, the case of the staff side was always admirably presented to the court whenever he appeared. I was often greatly interested and amused by the psychological display when William John Brown, who had been educated in the Salmestone Elementary School, Margate, and the Sandwich Grammar School, and had begun his career as a boy clerk in the Civil Service, on one side, confronted a young Treasury official in the Administrative Class, educated at Winchester and New College, Oxford, on the other. Winchester and Oxford were often no match for Margate and Sandwich.

The work was often very interesting. The occupational diversity of the hundreds of thousands of persons—or is it millions?—employed by the state is amazing, and the arbitrators almost always had to get a detailed knowledge of the work of claimants in order to determine what should be a fair wage or salary for the work which they did. In the course of my service on the Tribunal I learned all about the hourly work of the cohort of women who clean out the government offices in Whitehall, foresters in the north of Scotland, the men who talk down aeroplanes at certain airports in fog, and a small and peculiar class of men in the secret service. In order to try the last case, the Tribunal had to be indoctrinated and de-indoctrinated (if that was the correct word) by an army officer. Though this infinite variety made the whole business much more interesting for the arbitrator, the longer I served on the Arbitration Tribunal, the more crazily irrational the whole industrial structure of the Civil Service appeared to me. The Service was divided into classes, e.g. administrative, executive, clerical, and there were a certain number of subclasses and grades. But throughout this enormous business each of the hundreds of government occupations had, on the face of it, a scale of pay and conditions of employment peculiar to itself. On the other hand, the Civil Service Clerical Association, the organized scientific workers, and the trade unions claimed that throughout government employment similar work should entitle the worker to similar pay, and the Treasury agreed. The consequence is that, if 502 workers in occupation A receive a 5 per cent. increase in the scale of their pay, 5,003 workers in occupation B, 47 in occupation C, and 50,004 in occupation D will probably claim a similar increase on the ground that their work is similar to that of occupation A. I should guess that about 80 per cent. of the claims which I had to try in my 17 years on the Tribunal were of this kind. This meant that one had to listen to a detailed exposition of the exact nature of the work in each occupation and then judge whether they were sufficiently similar to justify a similar scale of pay. And one knew that if one assimilated the scale of, say, D to that of A, it was quite probable that the increase of the scale of D would provoke a claim from occupation E and occupation F to a similar increase.

The inevitable result of this system is an unending chain reaction of claims, a successful claim in one part of the Service setting off one or more claims somewhere else. The amount of wasted time and money in negotiations and arbitrations must be considerable. I am sure that much could be done to rationalize the structure of the Civil Service in this respect, though to do so would require courage and determination. If, as was proved again and again on the Tribunal, the work and conditions of occupations A to E, for instance, are so similar that the Tribunal has to give them the same scale of pay and conditions of employment, then those who work in the five occupations should all be recruited into one large government service Division, divided into Classes, each with its fixed scale of pay. This rationalization and reclassification of government employment ought to be done by a Special Commission. It might not be possible to apply the new classification and pay structure to people already in government employ, but it could be applied to all recruits after a certain date. I do not suppose that it would ever be possible to rationalize the whole government service, from top to bottom, in this way; there would always be some occupations so peculiar that their pay and conditions could not be assimilated to any other. But there would not be a great number, and the vast majority of government employees could equitably and efficiently be recruited as, say, a member of Class IV of Division X in the Civil Service. If they were, the work of the Civil Service Arbitration Tribunal would dwindle.

My exit from the Tribunal after 17 years amused me. Though I was on the panel nominated by the staff, I always considered that, as a member of the Tribunal, I had to be completely impartial. And on the whole, I think, we all of us were pretty impartial. Our decisions were nearly always unanimous. On two or three occasions only I found myself in a minority of one; once the chairman and I were on one side, and the Treasury arbitrator on the other. The written statements of the Treasury case were nearly always admirable and the statement of claims by the Civil Service Clerical Association were usually good, but the claims from other bodies were sometimes set out badly. In 1954 one of the big unions put in a claim for a large number of government employees; the statement of claim put in by them seemed to me perfunctory, slovenly; it asked for an increase in the scale of wages, but did not even take the trouble to give the existing scale. When we three arbitrators met before going into the court room, I said that I thought the statement extremely bad and I suggested that the chairman might gently draw the attention of the claimants to the deficiencies.

The other two agreed and the chairman made the mildest of protests. The secretary of the union, who had come to conduct the case, was enraged and refused to continue before us —so we all trooped out astonished rather than dismayed. Arbitrators were appointed to the panels for four years, and hitherto when my four years' term had come to an end, I had always been renominated and reappointed. When shortly after this incident my four-year term came to an end, I was not renominated.

When one comes to the practice of politics, anyone writing about his life in the years 1924-1939 must answer the crucial question: 'What did you do in the General Strike?' Of all public events in home politics during my lifetime, the General Strike was the most painful, the most horrifying. The treatment of the miners by the government after the Sankey Commission was disgracefully dishonest. If ever there has been right on the side of the workers in an industrial dispute, it was on the side of the miners in the years after the war; if ever a strike and a general strike were justified, it was in 1926. The actions of the mine-owners and of the government seemed to me appalling and when the General Strike came, I was entirely on the side of the workers. There was, of course, really nothing one could do, and one watched appalled the incompetence of those who had called and were conducting the strike. Then when the failure of the strike was inevitable, I was rung up one morning by R. H. Tawney, who asked me to come round and see him in Mecklenburgh Square. When I got there, he told me that he was going to try to get as many well-known people as possible to sign a statement publicly calling upon the government to see that there was no victimization when the strike was over. He asked me whether I would be responsible for collecting the signatures of as many prominent writers and artists as possible. I agreed and for the next few days organized a company of young people who bicycled round London collecting signatures. It was the kind of job which I find depressing, because I cannot really believe in the efficacy of what I am doing. However, we worked hectically, and only one person refused when asked to sign. This was John Galsworthy. The young woman who bicycled up to Hampstead and received a pretty curt refusal became Principal of Somerville College and a D.B.E.

In the general practice of politics my main activities were in the Fabian Society and the Labour Party. For many years I was elected a member of the Executive Committee of the Fabian Society, for ten years I was chairman of the Fabian International Bureau, and for even longer was a member of the Colonial Bureau. I do not like to think of the innumerable hours of my life which I have spent on this kind of work. It was mainly committee work. I do not enjoy committees and I am not a good committee man unless I am chairman or secretary; as an ordinary member, I tend either to become exasperated by what seems to me inefficiency and waste of time or to sink into a coma from which at long intervals I rouse myself to sudden, irritated energy. My attitude to the Fabian Society was, I think, always slightly ambivalent. When I first knew it, it was pre-eminently a creation of Sidney and Beatrice Webb. They meant it to be and they made it an instrument for the political education of the labour movement and ultimately of the Labour Party. By what they understood as 'research', by committees, reports, pamphlets, books, and conferences, the Society fed the labour movement with facts and theories, with ideas and policies. I was mainly interested politically in international affairs and in colonial and imperial problems. When, at the end of the 1914 war, Asquith, Lloyd George, and the ruthless logic of history had sterilized and emasculated liberalism and had irrevocably destroyed the Liberal Party, and when

Labour was just emerging as the only political alternative to conservatism and the Tory Party, both among the leaders and the rank and file of the labour movement there was a profound and almost universal ignorance of international and imperial facts and problems. The rank and file were predominantly working class and trade unionist, and were naturally concerned with the industrial and economic aspects of politics of which their knowledge and interests alone made them acutely aware. Even their middle-class leaders and instructors, like the Webbs, before the 1914 war ignored and were ignorant of international and imperial problems. As I have recorded elsewhere, when I first knew Sidney, if an international question cropped up, he would say: 'It's not my subject', and that seemed to mean that it was no business of his and could be left to some other expert. After I had written *International Government* and *Empire and Commerce in Africa* for the Fabian Society, the Webbs treated me more or less as the Fabian and Labour Party 'expert' in international and imperial questions. In the Society Sidney turned over to me or consulted me about such questions and he had me appointed secretary of the Labour Party Advisory Committees on International and Imperial Affairs.

My work in the Society helped my work in the Labour Party and vice versa, though I always felt that the Advisory Committee work was potentially the more important, since it brought one directly in contact with the Executive Committee of the Labour Party and the Parliamentary Party. During the 1920s I did a good deal of work for the Fabian Society, but I did not take much part in its internal politics, though I watched them from a distance with some amusement. At the beginning of the decade, as I have said, the Society was still mainly the creation of the Webbs, who still, spiritually and materially, moved on the face of its waters. But the younger generation, in the person of G. D. H. Cole

and his wife Margaret, were already knocking at the door—
and there was no gentleness or consideration when Douglas
and Margaret Cole knocked on any door. I have always
thought that the way in which Sidney and Beatrice behaved
when they heard that ominous knocking was a perfect
example of how age should face with understanding and
dignity the menace of the younger generation. When Douglas married Margaret in 1918, both were under 30 years of
age; he was an extraordinarily able product of Oxford University and she of Cambridge University. They did not
suffer either fools or the opposite of fools gladly; they knew
exactly what they wanted in life and in the Fabian Society,
and they were determined to get it with the ruthlessness and
arrogance of vigorous youth confronted by distinguished
and static age. There was no gentleness in their opposition
to the Webbs, but Sidney and Beatrice treated them and
their views with the greatest consideration and never showed
the slightest sign of resentment.

But this kind of thing was not very good for the work of
the Fabian Society and between 1920 and 1930 it gradually
went downhill. Indeed, it had got into such a miserable state
that in 1931 the more active members hived off and started
the New Fabian Research Bureau. Douglas Cole was honorary secretary, Clem Attlee, chairman, and Hugh Gaitskell,
assistant secretary. I was a member of the Executive Committee and ran the International Section. We did a considerable amount of work, producing pamphlets and reports. But
after nine years, negotiations opened with the Fabian Society
and we hived back again. The hiving back really meant
that we, the New Fabian Researchers, took control of the
Fabian Society. As far as I was concerned, I was elected a
member of the Executive Committee and served on it for
many years and I was also chairman of the International
Bureau.

I did a good deal of work for the Fabian Society, but much more for the Labour Party. For nearly 30 years I was secretary of the two Advisory Committees each of which met in the House of Commons on alternate Wednesdays. The present General Secretary of the Labour Party (1966) has given me an opportunity of examining the reports which these two committees made to the Executive Committee and the Parliamentary Party. I knew that we had done a great deal, but I must say that I was amazed at the quantity and scope of our work. We simply bombarded the leaders of the Party and the active politicians with reports, briefs, recommendations, policies covering every aspect, question, or problem of international and imperial politics. As a civil servant, an editor of journals dealing with the theory and practice of politics, and a publisher who specialized to some extent in the publication of political books, I have learned— at the cost of infinite boredom and much mental torture— a great deal about political writers and historians, and how they deal with their subjects. Nearly everyone thinks that he can think, and most people today think that they can write what they think; in fact, the ability to think even among professional thinkers, and to write even among professional writers, is extremely rare; and I should put most political writers and historians at the bottom of the class. It is therefore extraordinary to find that the quality of the reports of these committees is in general extremely high; most of them are succinct and to the point, written clearly by experts for ordinary people, full of facts, expressing views and proposing policies with a sense of sober responsibility.

I will deal first with the imperial committee. Today, in the year 1966, imperialism and colonialism are among the dirtiest of all dirty political words. That was not the case 47 years ago at the end of the 1914 war. The British and French Empires were still going strong and still adding to their territories, either unashamedly or, rather shamefacedly and dishonestly, by the newly invented Mandate system, which some people recognized as a euphemism for imperialism. The vast majority of Frenchmen and Britons were extremely proud of their empires and considered that it was self-evident that it was for the benefit of the world as well as in their own interests that they ruled directly or dominated indirectly the greater part of Asia and Africa. It was still widely accepted that God had so ordered the world that both individuals and states benefited everyone, including their victims, by making the maximum profit for themselves in every way, everywhere and everywhen. In the second volume of my autobiography, *Growing*, I recorded my experience as an imperialist empire builder for seven years in the Ceylon Civil Service and how gradually it made me dislike imperialism with its relation of dominant to subject peoples.[1] It was one of the main reasons why in 1911 I decided to resign from the Civil Service. The 1914 war and my work on *International Government* and *Empire and Commerce in Africa* increased my dislike of the imperialist system. It seemed to me certain that the revolt of the subject peoples— 'peoples not yet able to stand by themselves under the strenuous conditions of the modern world', as the Covenant of the League describes them—which had begun in Japan and was now rumbling in China, India, Ceylon, and the Near East, would spread through Asia and would soon reach Africa. In the modern world this was one of the most menacing political problems, and the world's peace and prosperity in the future depended upon accelerating the transfer of power from the imperialist states to their subject peoples in the ramshackle territories called colonies, dependencies, protectorates, and spheres of influence throughout Asia and Africa.

[1] *Growing*, pp. 158, 247-248, 251.

The importance of the Labour Party Advisory Committee on Imperial Affairs consisted for me in the fact that it enabled me, as secretary, to try to get the party and its leaders to understand the complications and urgency of what was happening in remote places and among strange peoples about whom they were profoundly and complacently ignorant. When it came down to day-to-day practical politics in the 'twenties and 'thirties of the twentieth century, there were in this field two main questions which it was essential to deal with: first, the demand for self-government in India and its repercussions in Burma and Ceylon; secondly, the methods of government and economic exploitation in the British African 'possessions'. Between 1918 and 1939 the Advisory Committee did an immense amount of work upon these two questions and I was very lucky to have several members who not only agreed with my view of their importance but had a profound and practical knowledge of them. I propose to say something about what we aimed at and accomplished in each case.

First as regards India. Between 1924 and 1931 we sent 23 reports and recommendations on India to the Executive Committee and Parliamentary Party. We had on our committee four members with a wide and detailed knowledge of India, Major Graham Pole, H. S. L. Polak, G. T. Garratt, and above all Sir John Maynard. They closely followed events and we relied greatly upon them and their intimate knowledge of Indian conditions. Maynard, who became committee chairman, was as I wrote in *Beginning Again* (p. 228), a very remarkable man. After a distinguished career in the Indian Civil Service he retired in 1927. He was for five years Member of the Executive Council of Governor Punjab. When he joined the Labour Party and the Advisory Committee he was well over 60; at that age most people have lost any ability, if they ever had it, of admitting into their

mind even the shadow of a new idea, and, after over 40 years
as a civil servant in India, he ought to have developed a sun-
baked shell of impermeable conservatism. When one first
met him, one might easily mistake him for the typical top-
grade British civil servant, neat, precise, reserved, reticent,
with an observant, suspicious, and slightly ironical gleam in
his eye. He may have been to some extent all of this, and
yet at the same time he was the exact opposite. For he was one
of the most open-minded and liberal-minded men I have ever
known—he was also personally one of the nicest—and to
add to all his other gifts, after a life spent in administration,
at the age of 75 he wrote one of the best books on Soviet
Russia ever published, *The Russian Peasant and Other
Studies*. Graham Pole, a Labour M.P. from 1929 to 1931,
had a considerable knowledge of India and was an indefatig-
able member of the Advisory Committee; he was convinced
that immediate steps should be taken to give self-government to India. Polak had the same equipment and outlook
as Graham Pole, and he was an intimate friend of Gandhi.

Among the Labour leaders inside and outside Parliament
there was very little knowledge or understanding of the
Indian situation. In the highest regions Ramsay MacDonald
posed as the expert in chief, having been a member of the
Royal Commission on Indian Public Services in 1912 and
having written two books on the 'awakening' and government of India. He was not entirely ignorant of the situation
and took an interest in what we were doing on the committee,
occasionally speaking to me or writing to me about our
memoranda and recommendations. He professed to be, as
usual, upon what I considered to be the side of the angels—
in favour of self-government, but, as usual, he was entirely
untrustworthy—I always mistrusted Ramsay, and particularly when he brought me the gift of agreeing with me. It was
almost the inevitable sign that he would find some reason

for doing the opposite of what he had agreed with you ought to be done. The only other front rank Labour leader who knew anything about India was Clem Attlee, but his interest in and knowledge of it only began in 1927, when he was a member of the Indian Statutory Commission. After becoming Secretary of State for the Dominions in 1942 and Prime Minister in 1945, he played a prominent part in the final events which brought independence and Dominion status to India and Pakistan in 1947.

The perpetual tragedy of history is that things are perpetually being done ten or twenty years too late. When in the last days and hours before the outbreak of the 1914 war, that monstrously unnecessary war, Grey and some of the other European statesmen were trying frantically to put forward proposals to stop it and the general staffs were taking frantic measures to make it inevitable, one of the Foreign Secretaries—I think it was the Austrian—to whom Grey had desperately telegraphed one more proposal which would have involved stopping Austrian mobilization, plaintively replied that events had once more outstripped and outdated the proposal. This outstripping and outdating of the proposals, policies, and acts of governments and statesmen by events is the perpetual story of human history. Over and over again the oncoming of some horrible and unnecessary historic catastrophe—some war or revolution or Hitlerian savagery—is visible and a voice is heard in the wilderness saying to the kings and presidents and prime ministers: 'If you want to prevent this catastrophe you must do X'. Then all the kings, presidents, and prime ministers, the governments, establishments, powers, and principalities shout in unison: 'This is the voice of Thersites, and Jack Cade, and Jacques Bonhomme, of Danton and Marat, of Bakunin and Karl Marx, of bloody revolutionaries, Bolsheviks, Left Wing intellectuals, and Utopians. We conservatives are the only

realists—it is fatal to alter anything except the buttons on a
uniform or what makes no matter—but to do X would be the
end of civilization.' Ten, twenty, fifty years later, when it is ten
years, twenty years, fifty years too late, when events have outstripped and outdated X which would have saved civilization,
then at last with civilization falling about their ears, the
realists grant X. This was the history of the French revolution,
of Home Rule and Ireland, of war and the League of Nations.

It was also the history of British governments and India
between 1920 and 1947. During the 1914 war the British
government had declared that it would co-operate with
Indians in order to establish self-government in India. The
White Paper, the Round Table Conference, and the India
Act of 1935 were the steps by which British conservative
and imperialist patriots sought honourably to dishonour
this promise. What they gave with one hand—niggardly
reforms—they took away with the other—the massacre at
Amritsar, the Rowlatt ordinances, the cat-and-mouse im-
prisonments and releases of Gandhi and Congress leaders.
The vicious circle of repression and sedition, sedition and
repression—the implacable legal violence of an alien government and the murderous, illegal violence of native terrorists
—established itself. At each stage the demands of Congress
for self-government and Dominion status were met by such
grudging and contemptible dollops of self-government that
any politically conscious Indian could only conclude that
once more the tragedy of freedom would have to be acted out
in India—the alien rulers would release their hold on the
subject people only if forced to do so by bloody violence.

Of course, no one can be sure of any What Might Have
Been in history. But I have no doubt that if British governments had been prepared in India to grant in 1900 what they
refused in 1900 but granted in 1920; or to grant in 1920
what they refused in 1920 but granted in 1940; or to grant

in 1940 what they refused in 1940 but granted in 1947—
then nine-tenths of the misery, hatred, and violence, the
imprisonings and terrorism, the murders, floggings, shoot-
ings, assassinations, even the racial massacres would have
been avoided; the transference of power might well have
been accomplished peacefully, even possibly without partition. At any rate all through the crucial years between 1920
and 1940 the Advisory Committee urged the Executive
Committee and the Parliamentary Party, and the Labour
Governments of 1924 and 1929, to do everything in their
power to meet the demands in India for self-government and
Dominion status. Our memoranda, written by Maynard,
Graham Pole, and Polak, continually explained to the Executive and Parliamentary Party the complicated and fluctuat-
ing situation, and the bewildering kaleidoscope of demands,
proposals, reports, White Papers, negotiations, Bills and
Acts. If we accomplished nothing else, we at least for the
first time did something to educate our masters; we got a
few Labour leaders to take an interest in and understand
what was happening in India. Our recommendations were
clear and consistent: at the crucial moments of the White
Paper, the Round Table Conference, and the Act of 1935,
we insisted that the government's offers or concessions were
totally inadequate. The Party accepted our recommendations
and came out publicly in favour of immediate steps to establish Dominion status.

All political parties make promises or announce policies
generously when they are not in power which they regret,
ignore, and repudiate when they obtain the power to carry
them out. In my 25 years' service as secretary of the Imperial
Advisory Committee I had splendid opportunities of often
seeing that Labour governments and politicians were not
immune from this change of heart or mind; in my own personal
involvement with the Indian question, I have three vivid re-

collections of particular incidents, and one is closely connected
with such a change of mind. It was during the war, when
Attlee was Deputy Prime Minister and Secretary of State
for Dominion Affairs in the Churchill Government. Something had occurred with regard to India which the Advisory
Committee considered of immense importance. The exact
point I do not now remember, but the Committee felt that
something should be done which the Labour Party by its
declared policy was morally and politically bound to stand
for. In view of Attlee's position and his knowledge of the
Indian problem and of the Party's declared policy since 1927,
the committee decided that Charles Buxton and I should go
and see him and put the case as the committee saw it and the
desirability of doing everything possible in this case to follow the lines of our declared policy. Attlee gave us an interview in No. 11 Downing Street. Charles and I sat on one
side of a table and the Deputy Prime Minister on the other
side. I was not an intimate friend of his, but I had worked
with him quite a bit, particularly on the New Fabian Research Bureau; Charles Buxton knew him much better. But
we had a very frigid reception. He listened to what we had
to say and then dismissed us. I felt that somehow we had
committed or were thought to have committed something
rather worse than a political indiscretion by bringing this
inconvenient reminder into the Holy of Holies of power. I
almost felt a slight sense of guilt as I slunk out of Downing
Street past the policeman who during the war examined your
credentials before allowing you to enter it.

My second vision of an incident with regard to the long-
drawn-out tragedy of India is very different. It was in 1931,
just after the Round Table Conference had ended. A Labour
M.P., James Horrabin, said that Gandhi wanted to meet a
few Labour people and discuss with them what his future
action should be; Horrabin asked me to come in the evening

of December 3 to his flat for this purpose. It was a curious party, consisting of Gandhi and some ten or fifteen Labour people—who we all were, I cannot now remember, but my recollection is that we were nearly all of us 'intellectuals', not first-line politicians. Everyone has seen photographs of that strange little figure, the Mahatma Gandhi, and he has been described again and again; what seemed to me remarkable about him was that, unlike most people, he was in life almost exactly like the photographs of him. He was, I suppose, one of the few 'great men'—if there are such people—that I have met. I do not think that, if I had met him in Piccadilly or Calcutta or Colombo, I would have recognized his greatness; but sitting with him in Horrabin's room, I could not fail to feel that he was a remarkable man. At first sight he presented to one a body which was slightly inhuman, slightly ridiculous. But the moment he began to talk, I got the impression of great complexity—strength, subtlety, humour, and at the same time an extraordinary sweetness of disposition.

Gandhi said that he was not going to talk much himself. He had asked us to come because he felt that the end of the Round Table Conference had left him personally in a difficult position and he was not at all clear what line he should follow when he got back to India. He wanted us each in turn to tell him how we saw the situation and what we thought his immediate course of action should be. Then one after the other round the room each said his piece—I cannot say that I found my piece very easy or illuminating. When we had all said our say, there followed one of the most brilliant intellectual pyrotechnic displays which I have ever listened to. Gandhi thanked us and said that it would greatly help him if his friend Harold Laski, who was one of us, would try to sum up the various lines of judgment and advice which had emerged. Harold then stood up in front of the fireplace and gave the most lucid, faultless summary of the complicated, diverse expositions of ten or fifteen people to which he had been listening in the previous hour and a half. He spoke for about 20 minutes; he gave a perfect sketch of the pattern into which the various statements and opinions logically composed themselves; he never hesitated for a word or a thought, and, as far as I could see, he never missed a point. There was a kind of beauty in his exposition, a flawless certainty and simplicity which one feels in some works of art. Harold's mind, when properly used, was a wonderful intellectual instrument, though as years went by he was inclined to take the easy way and misuse it both for thinking and writing.

My third vision connected with India is in Artillery Mansions in Victoria Street and another famous Indian, Shri Jawaharlal Nehru. It was in February 1936 that I received a letter from Nehru saying that he would very much like to have a talk with me: would I come in the afternoon of Monday and see him in Artillery Mansions, where he was staying? He had just been elected President of the All India Congress. I went to keep my appointment on a cold, grey, foggy, dripping, London February afternoon. I knew Artillery Mansions well, for at one time my sister, Bella, had a flat there. It fills me with despair whenever I see it— even on a bright spring day. It must rank among the greatest masterpieces of Victorian architectural ugliness; but it is not only horribly ugly, it is a monument of dark, gloomy inconvenience. It is one of the many human habitations over which I have always felt the inscription should be: 'Abandon hope all ye who enter here'.

It was no bright spring day when I went to see the future Prime Minister of India; it was, as I said, one of those dirty yellow, pea-soup, dripping London February days which make the heart sink. My heart sank as I climbed the dark staircase to Nehru's flat. The door of the flat was open, and, as nothing happened when I rang the bell, I walked in. The doors of all the rooms were open and the rooms were barely furnished. Hesitatingly I looked into a room from which came the sound of conversation; in it were three chairs and a table, Nehru, and another Indian. Nehru told me to come in and sit down, and the quite private conversation between the two went on for five or ten minutes as if I had not been there. My experience in Ceylon had taught me that the European custom of domestic privacy is unknown to Asiatics. In Asia houses and rooms often have no doors, and, no matter what may be happening in a room, all kinds of extraneous people will probably be wandering aimlessly in and out of it.[1] Eventually the Indian left and I had a conversation for about half an hour with Nehru. We talked in the bare room with the door of the room open and the door of the flat open and, as it seemed to me, all the doors in the world open. This is not calculated to give one a sense of privacy and comfort on a cold February afternoon in London. I liked Nehru very much as a man; he was an intellectual of the intellectuals, on the surface gentle and sad. He

---

[1] I saw a curious example of this when I revisited Ceylon in 1960. I was taken by Mr Fernando, the head of the Civil Service, to see the Prime Minister, Mr Dahanayake. The previous Prime Minister had only a short time before been assassinated. Mr Dahanayake was living in a large house, the gates of which were guarded by two soldiers carrying rifles who stopped us and examined our papers and credentials with the greatest care before admitting us. But the large compound at the back of the house was separated from other compounds only by a low unguarded fence through which any would-be assassin could easily have come unseen. And I had a long conversation with the Prime Minister in an open, unguarded room and, while we talked, all kinds of people wandered vaguely in and out of it. I remember thinking at the time how very different the precautions would have been if I had gone to see the Prime Minister in Downing Street although it was nearly 150 years since a Prime Minister had been assassinated in England.

had great charm, and, though there was a congenital aloofness about him, I had no difficulty in talking to him. It was a rather strange and inconclusive conversation. I had thought and still think that he had intended to discuss politics and, in particular, imperial politics from the Labour angle with me. And in a vague way we did talk politics, the problems of India and Ceylon; but it was pretty vague and somehow or other we slipped into talking about life and books rather than the fall of empire and empires. After about half an hour I got up to go and Nehru asked me where I was going. I said that I was going to walk to the House of Commons to attend a Labour Party Advisory Committee there and he said that he would walk with me as he would like to go on with our conversation. When we got down into the extraordinary sort of gloomy well outside the front door of the Mansions, we found waiting a press photographer who wanted to take a photograph of Nehru. Nehru insisted upon my being included in the photograph, which is reproduced here in the picture section. The gloom of Artillery Mansions, of London on a February afternoon, of life in the middle of the twentieth century, as it weighed upon the future Prime Minister of India and the Honorary Secretary of the Labour Party Advisory Committee on Imperial Affairs, and on their dingy hats and overcoats, is observable in the photograph. We then walked on up Victoria Street to the House of Commons, talking about life and literature on the way. We parted at the door to the central lobby and I never saw Nehru again.

There was another subject which the Imperial Advisory Committee spent even more time on than that of India—imperialism in Africa, the government of British colonial territories and the treatment of their native inhabitants in Africa. In knowledge and experience of this kind of imperial question the committee was exceptionally strong. In Norman

Leys and MacGregor Ross we had two men who had spent many years of their lives as government servants in Kenya. Among other members were W. M. Macmillan, Director of Colonial Studies in the University of St Andrews, who had been educated and was subsequently a Professor in South Africa; Norman Bentwich, who had been in the Palestine Government and was Professor in Jerusalem; Lord Olivier, who had been in the Colonial Office, a colonial governor, and Secretary of State for India; T. Reid, who had been 26 years in the Ceylon Civil Service; Arthur Creech Jones, M.P., who became Secretary of State for the Colonies in 1946; Drummond Shiels, M.P., who was Parliamentary Under-Secretary of State for India in 1929 and Parliamentary Under-Secretary of State for the Colonies 1929-31. And in the chair we had Charles Roden Buxton, who came of a family which, ever since the abolition of the slave trade, had a hereditary interest and concern in the protection of the rights of subject peoples in the colonial empires.

As soon as the committee got to work I began to put before it memoranda on the government of the colonial empire and on its future. It was soon decided that we should produce for the Executive Committee a detailed report on the political and economic conditions and government in the British African colonial territories, together with, if possible, a detailed long-term policy which we could recommend for adoption by the party. In the next few years we spent a great deal of time and thought on this. As we gradually worked out a policy, the drafting of it was left to Buxton and me. The final report was a formidable document, but it was adopted in its entirety by the Executive Committee and published in a pamphlet under the title *The Empire in Africa: Labour's Policy*.

Rereading the pamphlet today, it seems to me remarkable that such a document should have been produced over 40 years ago and that the principles and policy in it should have been adopted by a political party which was on the point of being returned to power and therefore responsible for the government of the empire. The pamphlet dealt in detail with three subjects: (1) land and labour in our African possessions, (2) government and self-government, (3) education. It pointed out that Britain was pursuing two completely different and contradictory administrative policies in her east- and west-coast African possessions. On the west coast the policy was to preserve native rights in land, prevent its sale to Europeans, and promote a native community of agriculturalists and the growth of native industries; on the east coast the policy was to sell or lease immense areas of land to European syndicates or individuals, to help them to develop the country through 'hired' or forced native labour, and to confine the native population not working for Europeans to 'reserves'. Labour maintained that the east-coast system had deplorable results and that the right policy for the future was to treat land as the property of native communities so that there should be no economic exploitation of the native by the European, and the native should be given the opportunity of developing the economic resources of the land as a free man and for the benefit of the native community. As regards self-government, the declared policy was ultimately the establishment of native representation on Legislative Councils and the gradual transfer of responsible government to these Councils. In order to train Africans to govern themselves, the government must educate them for self-government by making primary education accessible for all African children, by the provision of training colleges for teachers, technical colleges, universities, and experimental and model farms.

When our private or public world is overwhelmed by deserved or undeserved misfortune, no one is more silly or enfuriating than the self-satisfied person who says to us: 'I told you so'. I do not wish to seem to be saying here: 'We told you so'. But nothing is more important than that people should realize that the inveterate political conservatism of human beings—and pre-eminently of the ruling castes and classes—has produced an unending series of unnecessary historical horrors and disasters, ever since the Lord began it by trying to prevent Adam and Eve from learning the truth about the badly devised universe and world which He ill-advisedly had just created. I have lived nearly half a century since the end of the 1914 war, watching go by what must have been probably the most senselessly horrible 50 years in human history. When a hundred years hence the historian can calmly and objectively survey what we have seen and suffered, he will almost certainly conclude that fundamentally the most crucial events of the period were the revolt in Asia and Africa against European imperialism and the liquidation of empires. It has been a process of slow torture to millions of ignorant and innocent human beings— misery and massacre in Asia from India to China, Indo-China, Korea, Vietnam, and Indonesia; anarchy, massacre, and misery in Africa from Algiers and Mau Mau in Kenya to the Congo and Rhodesia. And the objective historian in 2066 will also, I feel sure, conclude that a very great deal of this misery and massacre would have been avoided if the imperialist powers had not blindly and doggedly resisted the demands of the subject peoples, but had carried out their own principles and promises by educating and leading them to independence.

At any rate for 20 years the Advisory Committee persistently pressed upon the Labour Party Executive the necessity for forestalling events by preparing and promoting self-government throughout Britain's colonial empire. We were not concerned with pious promises or generalizations.

We continually put before the Executive detailed practical proposals to meet the actual situation throughout Asia and Africa, whether in India or the Far East, in Kenya or Rhodesia. Again and again the Executive accepted our recommendations and publicly announced them to be the Party's policy either in official pamphlets, drafted by us, or by resolutions passed at annual conferences.

I have said above that it is characteristic of politicians and political parties to announce policies and make promises when in opposition and to forget or repudiate them when in power. Between 1919 and 1939 there were two Labour governments; the first lasted nine months, the second two years, and neither commanded a majority in the House of Commons. Obviously this meant that the time was too short, the programme too crowded, and the voting strength in the House too small for the government to take any major steps in the carrying out of Labour's colonial policy. Nevertheless, the record of Ramsay MacDonald's government in Asia and Africa seemed to me and a good many other people very disappointing by failing to carry out its promises in cases where it could and should have done so. An incident connected with the Advisory Committee, which I have already referred to, will show what I mean. In 1930 in Ramsay's second government Sidney Webb was Secretary of State for the Colonies and Drummond Shiels, a Scottish Labour M.P., was Under-Secretary. Sidney was in politics curiously ambivalent; he must have been born half a little conservative and half a little liberal. He was a progressive, even a revolutionary, in some economic and social spheres; where the British Empire was concerned, he was a common or garden imperialist conservative. Shiels, a medical man by profession, had been an assiduous member of the Advisory Committee and had, I think, learnt a great deal from it. There was nothing of the wild revolutionary about him; he was a hardheaded, liberal-minded, unsentimental Scot, and he was a convinced believer in the necessity for putting into practice the colonial policies worked out by the Advisory Committee and adopted by the Party. He used from time to time to ask me to come to the Colonial Office and discuss things with him. He was dismayed by Sidney's conservatism and his masterly inactivity whenever an opportunity arose to do something different from what Conservative governments and the Colonial Office Civil Servants had endorsed as safe, sound, and 'progressive' for the last half-century.

Towards the end of 1930 the Advisory Committee decided that an opportunity had occurred for trying to get Sidney to implement a small part of the Labour Party's policy. In those days the budgets of Crown Colonies had to be 'laid on the table' of the House of Commons and approved by the Secretary of State. I kept my eye on them so that any import-ant point with regard to a budget might be discussed by the Advisory Committee. I brought the Kenya 1930 budget before the Committee when it was laid on the table of the House. The Committee considered that the proposed ex-penditure on education and communications was grossly unfair to the natives. The amount to be voted for the educa-tion of white children was enormously higher per head than that for the education of African children; the proposed expenditure on roads to serve the white settlers' estates was far higher than that proposed for roads serving the native reserves. On the other hand, the taxation of Africans was proportionally much more severe than that of the settlers.

The Committee decided that Charles Buxton and I should ask Sidney Webb to see us, and that we should point out to him that this discrimination against the African was absol-utely opposed to the Labour Government's policy with regard to the education of Africans and promotion of African agriculture, and that the Secretary of State for the Colonies should insist upon a revision of the budget. Sidney, who was now Lord Passfield, for some strange reason asked us to meet him at the House of Lords instead of at the Colonial Office. We had an absurd meeting with Sidney in the red and gold Chamber of the House of Lords, which was, of course, completely empty except for the tiny Secretary of State for the Colonies and the humble chairman and secretary of the Advisory Committee sitting one on either side of him. We got, as I had expected, nothing out of Sidney, who was an expert negotiator and had at his fingers' ends all the arguments of all the men of action for always doing nothing.

This kind of thing, which often happened, made one wonder whether the immense amount of work which these Advisory Committees did was of any use at all. From time to time, off and on, one or other of the most active members would come to me and say that they had had enough of it, that we did an immense amount of work, pouring out reports and memoranda which the Executive Committee accepted, which became 'Party policy', and were then never heard of again. I tried to mollify and console them by recommending to them the rule by which I regulated my life and its hopes and fears: 'Blessed is he who expects nothing, for unexpectedly he may somewhere, some time achieve something'. The two Advisory Committees of which I was secretary did occasionally achieve something, though nothing commensurate with the amount of work we did. We spread through the Labour Party, and to some extent beyond it, some knowledge of the relations between the imperialist powers and the subject peoples of Asia and Africa, and even some realization of the urgent need for revolutionary reform so that there would be a rapid and orderly transition from imperialist rule to self-government. And within the Labour Party— particularly in the Parliamentary Party—the committees did a useful educational job. When Drummond Shiels became Under-Secretary of State for the Colonies in 1929 and Arthur Creech Jones became Secretary of State for the Colonies in 1946, they were far better equipped for their jobs than most M.P.s who get ministerial office, and they both of them did very good work in the Colonial Office. They would, I am sure, have agreed that they had learnt a great deal from the Advisory Committee; the same is true of many rank and file Labour M.P.s who never reached the rank of Minister.

Much of what I have said about the Imperial Advisory Committee applied to the International Advisory Committee. As I have said before, in 1919 the ignorance of foreign affairs in the Labour movement was almost as deep and widespread as the ignorance of imperial affairs. Like its sister committee, the International Committee performed a useful educational function in the Parliamentary Party, and even outside it. It was, perhaps, in membership rather stronger than the Imperial Committee. Until 1937 Charles Buxton was chairman, and two 'experts' in foreign affairs, Philip Noel-Baker and Will Arnold-Forster, did a great deal of work on it. In the 1920s both Ramsay MacDonald and Arthur Henderson kept in close touch with our work and the upper political stratum in the Party always treated the International Committee with rather more respect than the Imperial Committee. For instance, in June 1937, Ernest Bevin attended a meeting and developed to us the points which he had made in an important speech at Southport. There were 23 members present, and all, including Bevin, agreed that a memorandum should be sent to the Executive urging that a constructive peace policy should be put forward in the House of Commons by the front bench. The policy should be that the muddled pacts against aggression recently made by the government should be brought together into a uniform system, and it should be made clear that there would be an instant retort in event of any further act of flagrant aggression by Hitler.

In the terrible years between 1919 and 1939 everything in international affairs was dominated by the emergence of fascism in Europe and the menace of another war. To make up one's mind what seemed to be the right foreign policy and, where passions and prejudices became more and more violent, to keep cool and have the courage of one's convictions was a difficult, often an agonizing business. Even today it is difficult to write truthfully and objectively about those years and the part which one played in them, for the passions and prejudices persist and distort history.

I propose first to give an account of my own attitude during those 20 years and then describe the trend of opinion and policy, as I saw it, in the Advisory Committee and the Labour Party. Like most people on the Left who had some knowledge of European history and of international affairs, I thought that the Versailles Treaty, particularly in the reparation clauses, was punitively unjust to the German people, that it would therefore encourage militarism and desire for revenge in Germany. In this way it had sown the seeds of a second world war. I accepted Maynard's arguments in *The Economic Consequences of the Peace*. It seemed to me disastrous that, instead of supporting and encouraging a pacific, democratic Social-Democratic German government, France and to a lesser extent Britain did everything calculated to weaken and discredit it. The final folly, which played into the hands of the German nationalists and militarists, was the occupation of the Ruhr.

This attitude towards Germany and France was condemned at the time by many people as pro-German. It is still today condemned by some as pro-German and short-sighted; the subsequent history of Hitler, the Nazis, and the 1939 war shows, they say, that the Versailles Treaty was far from being too harsh; it was too mild; the Allies should have made a recrudescence of German militarism impossible by subjecting Germany to a modern version of the treatment by which 2,000 years ago Rome settled Carthage. The real question in this dispute and argument is: which was the cart and which the horse in the years 1919 to 1924? What we said was that, if you demand impossible sums in reparations and unjustly penalize Germany, you will cause economic chaos, you will get no reparations, you will encourage the revival of German militarism and a demand for a revision of the treaty. Here you have sown the seeds of a future war. At least the course of history followed exactly as we had prophesied.

From 1920 to about 1935 I thought that the international policy of the British Government, and therefore of the Labour Party, should be based on the League of Nations; the aim should be to build it up into a really efficient instrument of international government, a system for developing co-operation between states, the peaceful settlement of disputes, and collective security and defence against aggression. At every opportunity I put before the Advisory Committee proposals for implementing this policy. In 1927 Will Arnold-Forster and I drafted a Convention for Pacific Settlement which I put before the Advisory Committee; it was intended to close a gap in the League system as laid down in the Covenant. I received the following characteristic letter from Ramsay MacDonald:

House of Commons   January 17th, 1927
My dear Woolf,

I have been reading a very admirable memorandum put up by you and Arnold Forster regarding a Convention for Pacific Settlement. I think it is really a good piece of work, although one may see the possibility of filling certain detailed proposals, the idea and the general line laid down, seem to me to be excellent. I hope something will be done with it.

Yours very sincerely,

J. Ramsay MacDonald

The cryptic and ungrammatical second sentence is, I think, one of Ramsay's usual backdoors of escape which would enable him, if necessary, when the time came, to sabotage the damned thing with a clear conscience. I rather think that in fact that was precisely what he did do later on when Henderson put our draft before the League of Nations.[1]

All through the 1920s the Labour Party maintained this policy that the strengthening of the League and the collective security system was the only effective way of preventing another war. For 17 out of those 20 years a Conservative government was in power, first under Baldwin and then under Neville Chamberlain; neither of these statesmen believed in a League policy or attempted to use or develop the League as an instrument of peace as between the major powers. In this they were supported by the great majority of conservative politicians, though the curious incident of the Peace Ballot and the Hoare-Laval abortive agreement in 1935 makes it probable that a considerable majority of the rank and file conservatives disagreed with their leaders and would have supported a League policy. The two crucial tests of the League and collective security came in 1932 when Japan attacked China and in 1935 when Italy attacked Abyssinia. In both cases Baldwin with his Foreign Secretaries, Sir John Simon and Sir Samuel Hoare, contrived that the

[1] I think this document was what was called 'The General Act', that it was put by Henderson before the General Assembly of the League of Nations and adopted; and then Ramsay refused to allow Henderson to ratify it. Mr Philip Noel-Baker confirms this.

League's collective security system with full sanctions against the aggressors should not be operated.

By 1935 I had personally become convinced that Baldwin and the French statesmen who thought and acted as he did had finally destroyed the League as an instrument for deterring aggression and preventing war. The rise of Hitler to power, his withdrawal from the League, his adoption of compulsory military service, followed by his reoccupation of the demilitarized Rhineland showed the precariousness of the international situation and the necessity to take steps to meet the menace of war from Nazi Germany. I wrote several memoranda to the Advisory Committee urging that the new situation required a new policy: the League was to all intents and purposes dead and it was fatal to go on using it as a mumbled incantation against war; the only possibility of deterring Hitler and preventing war was for Britain and France to unite with those powers, including the U.S.S.R. if possible, who would be prepared to guarantee the small powers against attack by Hitler. I also pointed out that, if the Labour Party was going to support new security agreements against fascist or nazi aggression in place of the obligations under the League Covenant, 'mere negative opposition to a policy of rearmament would be sterile and ineffective'; if the Party really meant to commit itself to a policy of resisting any further acts of aggression by Hitler, then it committed itself to the corollary that Britain must make itself strong enough on land and sea and in the air to defeat Hitler.

These memoranda for the first time provoked a deep division of opinion in the Advisory Committee. But this was only part of a widespread disagreement, a profound, uneasy, often concealed ambivalence which for many years had permeated the Labour movement. There was within it a very strong pacifist element, derived in part from the tradi-

tional internationalism of the Labour and Socialist movements of the nineteenth century, and in part from the strong
Liberal contingent which, with the break up of the Liberal
Party after the war, had joined the Labour Party. To oppose
armaments in general and to vote against the Service esti-
mates in particular was traditional policy. This may or may
not have made sense when the pacifists and their parties were
opposing the jingo or imperialist policies of Conservative
governments which made the armaments necessary. But
when the pacifist Labour people (and surviving Liberals)
accepted the obligations of the League Covenant and its
collective security system, they were faced by an entirely
different situation. How could they agree to commit Britain
to join with other members of the League in resisting any act
of aggression by military means, if necessary, and at the same
time refuse to provide the armaments which alone could make
such military resistance feasible? There were some Labour
pacifists who, when confronted with this dilemma, logically
took the view that the obligation under the Covenant to use
force to resist aggression could only lead to world wars and
should be repudiated. But there was a far larger number who
never faced the dilemma and whose policy therefore contained a profound and dangerous inconsistency. The
dilemma and the disagreement were for years habitually and
discreetly ignored or glossed over. But as the menace of
Hitler and another war became more manifest, the divergence
of view within the Party rose to the surface. The show-down
came at the Labour Party Conference in Brighton in 1935.
George Lansbury had been Leader of the Parliamentary
Party since 1931; he was one of those sentimental, muddle-
headed, slightly Pecksniffian good men who mean so well in
theory and do so much harm in practice. He was a convinced
believer in the desirability of having the best of two contradictory worlds, of undertaking the obligation under the

League to resist aggression without providing the arms which would be required for the resistance. At the Conference Ernest Bevin, who took the view—with which I agreed—that, if you were going to fight against Hitler or any other aggressor, you must have arms with which to fight—rose in the pretty Regency Pavilion and made the most devastating attack upon the unfortunate Lansbury that I have ever listened to in a public meeting. As I said in *Beginning Again* (p. 221), he battered the poor man to political death—Lansbury afterwards resigned the leadership—and, although I was politically entirely on the side of Bevin in this controversy, I could not help shrinking from the almost indecent cruelty with which he destroyed the slightly lachrymose, self-righteous Lansbury.

The Advisory Committee was, as I said, divided, like the Party, on this question. There was a majority in favour of the League system of collective security and armaments adequate for resisting aggression. But there was a minority consisting of some who took the pacifist view and, with their eyes open, opposed rearmament and of some who, as it seemed to me, shut their eyes to the dilemma, inconsistently combining support of resistance to aggression with opposition to rearmament. In consequence there was never unanimity on the Committee for my memoranda; the Committee always decided to forward them to the Executive Committee without any positive recommendation. A (to me) sad result of this disagreement was Charles Buxton's resignation of the chairmanship of the Advisory Committee. He was essentially what Pericles, Aristotle, and Theophrastus would have called a 'good man' both in public and in private life. He was really mentally and emotionally a nineteenth-century non-conformist Liberal of the best type and therefore never completely at home in the twentieth-century Labour Party. A gentle man, he was on the side of civilization, hating violence of all kinds, regarding it as a first duty to devote oneself un-selfishly to the public good. It was characteristic of him that he joined the Quakers and tried to translate the ethics of the Friends into political terms. It was a curious trait in this kind of nineteenth-century Liberal often to develop a not altogether rational attachment to some foreign nation, nationality, or race. There were pro-Turks, pro-Americans, pro-Boers, pro-Bulgarians, and in the twentieth century pro-Germans. In the Balkan wars of 1912 and 1913 Charles became a pro-Bulgarian, and in 1914 he and his brother Noel went to the Balkans on a mission the object of which was to keep Bulgaria out of the war. A Turk tried to assassinate them and shot Charles through the lung. The Versailles Treaty made him what was called a pro-German in the grim years of peace, and I do not think that in the 1930s, when Hitler and the Nazis came to power, he could bring himself to face the facts and the terrible menace of war and barbar-ism from Germany. He necessarily took the extreme pacifist view, and, as the majority on the committee held the views which I did, he resigned the chairmanship. He and I had worked closely together on both Advisory Committees for many years, and it was sad to see him go, though our poli-tical disagreement made no difference to our personal friendship.

My views on this subject involved me in a curious incident in 1938. When Hitler invaded Austria in March, I was con-vinced that the last glimmer of hope of preventing war was drastic action on the part of Britain and France, and that this would require a dramatic change of policy by the Labour Party. The evening after the invasion I was at a meeting or party at which several Labour people were present—inevit-ably discussing the situation. I said that I thought that the Executive Committee and the Parliamentary Party ought to have a joint meeting and instruct the Leader of the

Parliamentary Party to make a formal public statement on their behalf as follows: The danger of further aggression by Germany and of war was so acute that the Labour Party considered that drastic action was necessary on the part of Britain and France to warn Hitler that any further aggression would be resisted; with a view to this the Party would be willing to enter a coalition government under Mr Winston Churchill pledged to forward this policy and would agree to an immediate introduction of conscription and rearmament. Over the week-end I was rung up by someone who had heard what I said; he told me that my arguments had convinced him and some others and they thought it important to try to get the Labour leaders to take action along the line which I had suggested—would I come and discuss what we might do with a few people on Tuesday? I was dining out on the Tuesday, but after dinner I left my party and went round to Wansborough's flat in Russell Square. There I found Douglas Jay, Tommy Balogh, and, I think, Hugh Gaitskell and Evan Durbin; there may have been one or two others. After some discussion it was decided that we should try to get hold of one or two of the leaders and induce them to put my proposition before a joint meeting of the Labour Party and T.U.C. which was to be held later in the week. One of us—I forget who it was—undertook to see A. V. Alexander, who seemed to be one of the most likely leaders to put forward the policy, and I agreed to talk to Phil Noel-Baker. Alexander agreed with our arguments and proposal and half promised to come out publicly in favour of it at the conference if he could get support beforehand within the Parliamentary Party. I could not get hold of Phil, because he was at a meeting of the International in Paris. The whole thing fizzled out: Alexander could get little or no support and drew back. Nothing was done—and the herd, Europe and the world, continued downhill all the way under Hitler's direction and 'ran violently down a steep place' into war.

I must leave the subject of politics. The years 1930 to 1939 were horrible both publicly and privately. If one was middle aged or old and so had known at least a 'sort of a kind' of civilization, it was appalling impotently to watch the destruction of civilization by a powerful nation completely subservient to a gang of squalid, murderous hooligans. My nephew, Quentin Bell, who was 20 years old in 1930, has recently described what it felt like to be a young man alive in those years:

> Who but we can recall the horror of that period? Of course, it was not continuous: we had our gaieties, our moments of hope, of exhilaration, of triumph even. Nevertheless, they were years of mounting despair: unable to compound our internecine quarrels, unable to shake the complacency of a torpid nation, we saw the champions of tyranny, war and racial persecution winning a succession of ever easier victories. In those twilight days it was bloody to be alive and to be young was very hell.

To the middle aged, i.e. to those who were already going downhill all the way to old age or death, it was also often in those twilight days bloody to be alive and very hell to be no longer young. In 1938 I wrote a play, *The Hotel*, about the horrors of the twilight age of Europe, the kind of hush that fell upon us before the final catastrophe. It was published in 1939 and was republished by the Dial Press in America in 1963. I can best explain how I came to write it by quoting the introduction which I wrote for the American edition of 1963:

> It is a long time since I wrote this play, *The Hotel*, and it seems even longer. It was written and published in England just before the 1939 war, and Hitler and Stalin

and Mussolini—the nazis, communists, and fascists—
finally destroyed the world in which it was written. That,
after all, is what the play is about; what it prophesied has
happened. Looking about us today, we can say with
Stanovich: 'The ceiling's down; the clock's smashed; and
there's no door. There's no back to the hotel and no
boiler room, and the wind coming through is fair cruel. . . .
What a place! What a place!'

That, perhaps, is all that the author can say about the
play in an introduction to its publication, after more than
twenty years, in America. It was written in the tension of
those horrible years of Hitler's domination and of the
feeling that he would inevitably destroy civilization. There
is, however, one small point which I can add as author;
I had never written a play before I wrote *The Hotel*. But
for a long time I had wanted to write one in which the
scene would be the entrance hall of a hotel, with the
revolving door through which a string of heterogeneous
characters would have their entrances and their exits. It
is a scene in real life which always seems to me infinitely
dramatic. And then one day in 1938 I suddenly saw that
my hotel on the stage might be both realistic and symbolic,
the *Grand Hôtel du Paradis* which had become the *Grand
Hôtel de l'Univers et du Commerce*, with Peter Vajoff, the
proprietor, standing in front of the fire—and with bugs in
the beds.

I was 40 in 1920 and 60 in 1940. The twilight was in one's
private as well as in public life. Death is, according to Swinburne, one of the three things which 'make barren our lives';
'death', said Virginia in the last paragraph of *The Waves*,
'death is the enemy'. If one does not oneself die young, the
moment comes in one's life when death begins permanently
to loom in the background of life. Parents, brothers, and

sisters, who were parts of one's unconscious mind and memories, die; the intimate friends of one's youth die; our loves die. Each death as it comes, so inevitable of course, but always so unexpected and so outrageous, is like a blow on the head or the heart. Into each grave goes some tiny portion of oneself.

This erosion of life by death began for Virginia and me in the early 1930s and gathered momentum as we went downhill to war and her own death. It began on 21 January 1932 when Lytton Strachey died of cancer. This was the beginning of the end of what we used to call Old Bloomsbury. Lytton was perhaps the most individual person whom I have ever known. His father was a Strachey and his mother a Grant; he came, therefore, on both sides from one of those distinguished upper middle-class families of country gentlemen who in the nineteenth century found their professional and economic home in India or the army. The mixture of Strachey and Grant blood in Lytton's family produced remarkable results; I gave some account of it in *Sowing* (pp. 202-208). It consisted of ten sons and daughters, all of whom were extremely intelligent and many of them intellectually remarkable. Lytton was unquestionably the most brilliant. He had an extremely subtle and supple mind, with a tremendously quick flicker of wit and humour continually playing through his thought. Everything about him—his mind, body, voice, thought, wit, and humour—was individually his own, unlike that of anyone else. His conversation was entrancing, for his talk was profounder, wittier, more interesting, and original than his writing. This was one of the reasons why his books, brilliant and successful though they were, slightly disappointed the expectations of many who had known him as a young man of 20. He had a tremendous reputation among the intellectuals as an undergraduate at Cambridge and we thought that he might well become a great Voltairian historian or biographer. He never achieved that, though *Eminent Victorians* and *Queen Victoria* are much more remarkable than they are currently and momentarily judged to be, and they obviously had a considerable influence on biography and history in the 20 years which followed their publication.

Lytton's personal influence on his own generation and those which immediately followed it at Cambridge was also very great. His personality was so strong that he imposed it, intellectually and even physically, upon people, especially the young. You could tell who saw much of him, for they almost inevitably acquired the peculiar Strachey voice which had a marked rhythm and, in his case, a habit of rising from the depths in the bass to a falsetto squeak. Lytton repelled and exasperated some people, particularly the dyed in the wool, athletic, public school Englishman, with no (but O so much) nonsense about him. By public school standards he did not look right, speak right, or even act right, and, apart from such major vices, he had the lesser faults of arrogance and selfishness. He would therefore often exasperate even his most intimate friends—but only momentarily and superficially. Fundamentally he was an extremely affectionate person and had (in life and conversation, though not always in his books) a great purity of intellectual honesty and curiosity. That was why his death shocked and saddened us so painfully: it was the beginning of the end, for it meant that the spring had finally died out of our lives.

After Lytton's death Carrington tried unsuccessfully to commit suicide. It was clear that sooner or later she would try again. Ralph asked us to come down to Ham Spray and see her; he thought we might be able to do some good. On March 10 I drove down with Virginia in the morning. It was one of the most painful days I have ever slowly suffered. The day itself was incongruously lovely, sunny, sparkling.

I remember most vividly Carrington's great pale blue eyes and the look of dead pain in them. The house was very cold; she gave us lunch and tea and we talked and she talked quite frankly about Lytton and his ways and his friends. At first she seemed calm and cowed—'helpless, deserted', as Virginia said, 'like some small animal left'. There was a moment when she kissed Virginia and burst into tears, and said: 'There is nothing left for me to do. I did everything for Lytton. But I've failed in everything else. People say he was very selfish to me. But he gave me everything else. I was devoted to my father. I hated my mother. Lytton was like a father to me. He taught me everything I know. He read poetry and French to me.' We left after tea, and just before we got into the car Virginia said to her: 'Then you will come and see us next week—or not—just as you like?' And Carrington said: 'Yes, I will come, or not'. Next morning she shot herself.

Two years later Roger Fry died, as the result of a fall in his room. Roger belonged, of course, to an earlier Cambridge generation than we did—he was 14 years my elder, but he was an integral part of Old Bloomsbury and of our lives. I have tried to describe his character in *Beginning Again* (pp. 93-98) and I will not repeat myself here. From 1920 until his death 14 years later he was indeed, as I have said, part of our lives. Living in Bernard Street, just round the corner from Gordon Square where the Bells lived and Tavistock Square where we lived, he was in and of Bloomsbury. With his death again something was torn out of our lives.

On July 18, 1937, death struck again when Vanessa's son, Julian, was killed driving an ambulance in the Spanish civil war. The story of Julian's life and death has been told at length, and with great skill, sympathy, and understanding, by two Americans, Peter Stansky and William Abrahams, in their book *Journey to the Frontier*. It would be silly of me to try to do in a page what they have done so well and fully in so many. I saw him at close quarters grow from a rampageous, riproaring child into a very large, serious, gay, rampageous youth and finally man—he was 29 when he was killed. His mother and father, Vanessa Stephen and Clive Bell, were extraordinarily dissimilar in mind, temperament, and looks, and there was, I think, an unresolved discord in Julian's genes the effect of which could be traced in his character, mind, and life. I have never known any child make so much noise so cheerfully, and he never quite grew up: there was still something of the child, riproaring round the sitting-room in Gordon Square, in the young man of 25 driving a car or having a love affair. He was an extremely attractive and lovable person and highly intelligent, but, like all ebullient and erratic people, he could at moments be exasperating. Virginia was devoted to him and so was I. His death and the manner of it, a sign and symptom of the 1930s, made another terrible hole in our lives.

Finally—a very different death—just before the war, on July 2, 1939, my mother died. She was then an old woman of 87 or 88; but in many ways she never grew old. She still retained an intense interest and curiosity in all sorts of things and persons, and was physically very active. Being short and fat and impulsive, and unwilling to accommodate herself to the limitations and infirmities of old age, she was always tripping over a footstool in her room or the curb in the street and ending with a broken arm or leg. At the age of 87 she did this once too often, for the broken limb this time led to complications and she died in the London Clinic. I described her character at some length in *Sowing* (pp. 34-39); of her nine surviving children, four, in mind and body, were predominantly Woolfs, my father's family; two were predominantly de Jonghs, my mother's family; and two were half and half. I was very much my father's and very little my mother's son, and there were many sides of my character and mind which were unsympathetic to my mother; I had no patience with her invincible, optimistic sentimentality, and my un-sentimentality, which seemed to her hardness and harshness, distressed her. There was no quarrel or rift between us, and I always went to see her once a week or once a fortnight up to the day of her death—but, though she would never have admitted it even to herself, I was, I think, her least-loved child. But there is some primitive valve in our hearts, some primeval cell in our brains—handed down to us from our reptilian, piscine, or simian ancestors, perhaps—which makes us peculiarly, primordially sensitive to the mother's death. As the coffin is lowered into the grave, there is a second severance of the umbilical chord.

With my mother's death we reached the beginning of the second war and, therefore, the end of this volume. I will actually end it with a little scene which took place in the last months of peace. They were the most terrible months of my life, for, helplessly and hopelessly, one watched the inevit-able approach of war. One of the most horrible things at that time was to listen on the wireless to the speeches of Hitler, the savage and insane ravings of a vindictive underdog who suddenly saw himself to be all-powerful. We were in Rodmell during the late summer of 1939, and I used to listen to those ranting, raving speeches. One afternoon I was planting in the orchard under an apple-tree iris reticulata, those lovely violet flowers which, like the daffodils, 'come before the swallow dares and take the winds of March with beauty'. Suddenly I heard Virginia's voice calling to me from the sitting-room window: 'Hitler is making a speech'. I shouted back: 'I shan't come. I'm planting iris and they will be flowering long after he is dead.' Last March, 21 years after Hitler committed suicide in the bunker, a few of those violet flowers still flowered under the apple-tree in the orchard.

INDEX

INDEX

INDEX

INDEX

Books by Leonard Woolf
available in paperback editions
from Harcourt Brace Jovanovich, Inc.

SOWING: AN AUTOBIOGRAPHY OF THE
YEARS 1880 TO 1904

GROWING: AN AUTOBIOGRAPHY OF THE
YEARS 1904 TO 1911

BEGINNING AGAIN: AN AUTOBIOGRAPHY
OF THE YEARS 1911 TO 1918

DOWNHILL ALL THE WAY: AN AUTO-
BIOGRAPHY OF THE YEARS 1919 TO 1939

THE JOURNEY NOT THE ARRIVAL MATTERS:
AN AUTOBIOGRAPHY OF THE YEARS
1939 TO 1969